W9-CEN-762

AP® PHYSICS 1
CRASH COURSE®

By Amy Johnson, M.A.

Research & Education Association
Visit our website at: www.rea.com

Research & Education Association
61 Ethel Road West
Piscataway, New Jersey 08854
E-mail: info@rea.com

AP® PHYSICS 1 CRASH COURSE®

Printed in the United States of America

Library of Congress Control Number 2015949495

ISBN-13: 978-0-7386-1196-9
ISBN-10: 0-7386-1196-4

AP Physics 1
CRASH COURSE
TABLE of CONTENTS

INTRODUCTION

CONTENT REVIEW

PART III

TEST-TAKING STRATEGIES

ONLINE PRACTICE EXAM........... *www.rea.com/studycenter*

ABOUT THIS BOOK

REA's *AP Physics 1 Crash Course* is the first book of its kind for the last-minute studier or any AP student who wants a quick refresher on the course. The *Crash Course* is based upon a careful analysis of the AP Physics 1 Course Description outline and available official AP test questions.

Written by an AP Physics 1 expert, our easy-to-read format gives students a crash course in the basic knowledge of physics, including theories and techniques, concepts, and general principles. The targeted review chapters prepare students for the exam by focusing only on the topics tested on the algebra-based AP Physics 1 exam.

Part One discusses the keys for success and shows you strategies to help you build your overall point score. Part Two is an overview of the content that will be covered on the exam, including Newton's laws of motion and force, circular motion, momentum, rotation, work, energy and power, electrostatics and simple circuits, and more.

Part Three gives you general AP test-taking strategies and teaches you how to master the multiple-choice and free-response sections of the exam. A chapter on laboratory analysis techniques is also included.

No matter how or when you prepare for the AP Physics 1 exam, REA's *Crash Course* will show you how to study efficiently and strategically, so you'll be ready for the exam.

To check your test readiness for the AP Physics 1 exam, either before or after studying this *Crash Course*, take REA's **FREE online practice exam**. To access your practice exam, visit the online REA Study Center at *www.rea.com/studycenter* and follow the on-screen instructions. This true-to-format test features automatic scoring, detailed explanations of all answers, and diagnostic score reporting that will help you identify your strengths and weaknesses so you'll be ready on exam day!

Good luck on your AP Physics 1 exam!

ABOUT OUR AUTHOR

Amy Johnson holds a B.A. in Physics Teaching from Brigham Young University and an M.A. in Physics Education from Smith College. She currently serves as the Science Director for Mass Insight. She has been teaching physics for more than 10 years at both the high school and college levels.

As a teacher for Northampton (Mass.) High School, Ms. Johnson worked to expand the AP Physics program to include both AP Physics B and C. Apart from broadening the program, she also helped students achieve success in their physics education and preparation for college. Ms. Johnson has also taught Physics for Middle School Science Teachers at the University of Massachusetts Amherst. She is the recipient of the Harold Grinspoon New Teacher of the Year Award, as well as the National Math and Science Initiative Science Teacher of the Year Award.

AUTHOR ACKNOWLEDGMENT

I would like to express my great appreciation to my husband, Aaron, for his love and support throughout this project, and for his willingness to stay up late into the night reading drafts. I would also like to offer my special thanks to all my readers, Jenny, Jessica, Charlie, Owen, Heather, and Heath, for their invaluable insight and tireless patience. Without your help I couldn't have finished this book! Thanks also to my family for their continual support of this effort—your encouragement made all the difference.

ABOUT OUR TECHNICAL EDITOR

Meghan Bjork received her B.A. from Gustavus Adolphus College, St. Peter, Minnesota, and her M.A in Teaching in Science Education from the University of Iowa, Iowa City, Iowa. She has been teaching AP Physics, Honors Physics, and Earth Science for eight years.

ABOUT REA

Founded in 1959, Research & Education Association (REA) is dedicated to publishing the finest and most effective educational materials—including study guides and test preps—for students in middle school, high school, college, graduate school, and beyond.

Today, REA's wide-ranging catalog is a leading resource for students, teachers, and other professionals. Visit *www.rea.com* to see a complete listing of all our titles.

ACKNOWLEDGMENTS

We would like to thank Pam Weston, Publisher, for setting the quality standards for production integrity and managing the publication to completion; John Paul Cording, Vice President, Technology, for coordinating the design and development of the REA Study Center; Larry B. Kling, Vice President, Editorial, for overall direction; Diane Goldschmidt, Managing Editor, for coordinating development of this edition; Kathy Caratozzolo of Caragraphics for typesetting this edition; Linda Robbian for copyediting; Ellen Gong for proofreading; and Eve Grinnell, Graphic Artist, for design and file management.

PART I
INTRODUCTION

Keys for Success
on the AP Physics 1 Exam

Congratulations on your decision to take AP Physics 1. Taking an AP course, especially an AP science course, is a challenging endeavor. AP courses represent college-level classes that you take in high school, so they are supposed to be challenging. But with your hard work, the help of your teacher, and this test prep, you will be guided toward success on this exam and in college.

The information presented in this book is meant to prepare you for the AP Physics 1 exam, based on information currently available from the College Board. The exam will test your critical-thinking skills, your ability to explain and reason both qualitatively and quantitatively, and your ability to show that you have an enduring understanding of physics that will support future coursework in science.

Let's take a look at the exam!

1. The Structure of the Exam

The AP Physics 1 exam consists of both multiple-choice and free-response questions.

Section I—Multiple Choice—90 Minutes

- 45 multiple-choice questions with only one correct answer

- 5 multiple-choice questions with two correct answers (you must choose *both* correct answers to get credit for the question)

Section II—Free-Response Questions—90 Minutes

There are 5 free-response questions on this section of the exam. These include:

- 1 experimental design question
- 1 quantitative/qualitative translation
- 3 short-answer questions

Each of the two sections of the exam is worth 50 percent of your score. The multiple-choice section is scored electronically and you are not penalized for guessing.

2. The AP Physics 1 Outline

The AP Physics 1 Curriculum Framework published by the College Board does not specifically state the units or topics that will be taught in a typical AP Physics 1 course. The Framework is instead organized around seven "Big Ideas" that will reoccur as themes.

The College Board is no longer publishing a topical breakdown for the AP Physics 1 exam, nor is it providing percentages of the exam that relate to each topic. We do, however, know that the following seven topics will be on the AP Physics 1 exam:

> Kinematics
>
> Forces, Circular Motion and Gravity
>
> Work and Energy
>
> Impulse and Momentum
>
> Rotation, Torque, and Angular Momentum
>
> Mechanical Waves, Sound, and Simple Harmonic Motion
>
> Electrostatics and Circuits

AP Physics 1 will require less calculation and more written explanation than its predecessor, the AP Physics B exam. In fact, it will require far more explanation than any other standardized physics exam. There will be fewer topics and less math, but it will require you to have a deeper conceptual understanding of the physics topics. AP Physics 1 is not about getting the right number, but about knowing (and being able to explain) what the numbers mean. You have to remember that in physics class, every number has a real meaning and represents real phenomena!

3. Types of questions on the exam

The AP Physics 1 exam contains several types of questions:

a. Descriptive problems

A descriptive problem is similar to what you've seen in your AP Physics 1 class and in your textbook. A descriptive problem sets up a situation and then asks you to conceptually discuss the situation.

Example:

A lightbulb is connected to a battery. A second identical lightbulb is connected in series with the first. Which of the following happens to the current and the resistance in the circuit?

(A)   Both increase

(B)   Both decrease

(C)   The current increases since the resistance decreases.

(D)   The current decreases since resistance increases.

b. Calculation problems

Calculation problems should be familiar to you from your class and textbook. These are straightforward calculations based on your knowledge of the situation and the equations.

Example:

A 5 kg box is pulled vertically at a constant speed of 5m/s. If the box is pulled for 3 seconds, the power developed during this time is

(A)   25 watts

(B)   75 watts

(C)   83 watts

(D)   250 watts

c. Ranking tasks

Ranking tasks present you with several pictures of similar objects or situations, and you are asked to rank them based on some unknown quantity.

Example:

Three softballs are thrown at different angles and at different speeds as shown below. Rank the balls on the basis of their acceleration at the top of their trajectory.

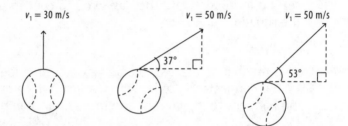

$v_1 = 30$ m/s $v_1 = 50$ m/s $v_1 = 50$ m/s

37°

53°

(A) $1 > 2 > 3$

(B) $3 > 2 > 1$

(C) $1 = 2 = 3$

(D) Accelerations are all zero at the top of the trajectory.

d. Semi-quantitative reasoning

Semi-quantitative reasoning questions ask you to solve problems that involve some algebra using only symbols. In a semi-quantitative reasoning problem, you might be asked to figure out how much further a car would skid before coming to a stop if it were going twice as fast on the same road. You'll have to do some algebra to figure out what the relationship is between the initial velocity of the car and the skid distance for the car, and then see how increasing the initial velocity by a factor of two changes the skid distance. If you struggle with this kind of question, keep practicing! (It *is* okay to plug in numbers to test if you get stuck!)

Example:

A box of mass *m* slides off the edge of a horizontal table at speed *v* and lands a distance *d* from the edge of the table. A second identical box leaves the table at *3v* and lands at:

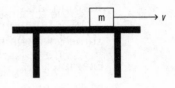

(A) *d*

(B) $\dfrac{d}{3}$

(C) 3*d*

(D) 9*d*

e. Experimental design

Experimental design questions set up a situation where an experimental procedure is needed to find a result. The answer choices will provide different ways to set up the required experiment. You must choose the experimental procedure that will get you the desired result. The best way to practice this kind of question is to get your hands on equipment and do lots of lab experiments yourself. By doing so, you'll have an intuitive idea as to what will work and what won't work.

Example:

An object of mass *m* is attached to a string of length *l* and is pulled back so that it is a height *h* above the lowest point. It is desired to determine what the speed *v* at the lowest point will be if the object is instead released from *2h*. Which of the following procedures would accomplish that determination?

(A) Position a motion detector so that it will catch the entire motion of the object. On the velocity vs. time graph output by the detector, look at the maximum vertical axis value which shows the highest speed attained by the object.

(B) Place a photogate where the object will pass through it at the lowest point. Measure the time for the object to pass through the photogate. Divide the distance the object swings by that time.

(C) Position a motion detector under the lowest position of the object. Divide the distance traveled by the object by the time recorded by the motion detector.

(D) Place a motion detector under the lowest position of the object. Find the area under the curve of the position vs. time output by the detector.

f. Experimental analysis

Experimental analysis questions ask you to decide the best way to present the results of an experiment. You will often be given a data set and asked to analyze, or find a way to graph the given quantities, to find a given result.

Example:

Students conduct an experiment to calculate the acceleration due to gravity using a pendulum. The length of the pendulum is varied and the period of oscillation is measured. Which of the following will produce a slope that can help the students find "g"?

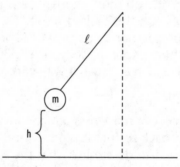

(A) T vs ℓ

(B) T^2 vs. ℓ

(C) T vs. ℓ^2

(D) $\sqrt{T}$ vs. ℓ

g. Multiple-correct questions

Multiple-correct questions are new to the AP Physics 1 exam and will be the last 5 questions on the multiple-choice section (Questions 46-50). For a multiple-correct question, two of the four answers will be correct and you will have to identify *both* correct answers to get credit for the question. There will not be partial credit given on this section.

Example:

Which of the following two adjustments would increase the resistance of a long wire?

(A)   Decrease the cross sectional area

(B)   Increase the length of the wire

(C)   Increase the voltage across the resistor

(D)   Decrease the current across the resistor

4. Basic Test-Taking Strategies

The best way to prepare for the AP Physics 1 exam is to pay attention and work hard in your AP Physics class at school. Take full advantage of your time in class and at home during the year.

To prepare for the exam, use this book and take REA's online practice exam. The practice exam will help you become comfortable with the format of the test and the types of questions that will be asked. If you need more practice tests, you can check the College Board website (*www.collegeboard.org*) for their released sample questions. The more you practice, the better you will do!

Read all the test questions carefully. There is a lot more reading on this exam than on other standardized exams. Take your time and focus on the question. The College Board knows that these questions take more time to read and answer, and they have given you that time, so don't rush!

Although you *can* use a calculator on both sections of the exam, you shouldn't need to use it more than a few times throughout the entire exam. Even on questions where "calculation" is required, you'll often be asked to make an estimate, or the numbers will be simple enough for you to solve the problem in your head. So, start weaning yourself from your calculator now. Practice using diagrams and graphs. Practice writing the equation and deciding what you could solve for without actually solving, and use your calculator only as a last resort!

You will also have a table of information and an equation sheet that you may use on both the free-response and multiple-choice sections. *But,* just like your calculator, you shouldn't need to use the equation sheet more than a few times. Now that calculation is a far less significant portion of the AP Physics 1 exam, the equation sheet will rarely be useful. If you do use it, use it carefully and wisely! Don't go fishing for an equation to use. (You may end up looking for an equation for power and find $p=mv$, where the p means momentum, not power.)

There is no penalty for guessing on the exam, so don't leave any question blank. Eliminate answer choices you know are wrong and then make your best guess. If you have time, go back and look over the questions that were difficult for you.

AP Physics 1 is not a broad course. It covers three basic topics: mechanics, electricity, and waves. The majority of the exam will be focused on mechanics. Keep in mind that AP Physics 1 is not a math course. You will have to solve equations with a single variable, calculate and understand the meaning of the slope and area under a graph, use the basic trigonometry functions of sine, cosine and tangent, and solve problems. But you also need to explain: *how* you solved the problem, *what* concepts you used, and *why* those concepts can be applied. (You can practice this all year as you work with the questions your teacher assigns. When you finish with a numerical answer, try writing a "why" paragraph.) Any numerical question in physics can be tested in the laboratory, and you must be able to articulate how this would be done.

5. What Should I Bring to the Exam?

1. Pencils and extra erasers

2. Black or blue pens for the free-response section

3. Calculator

4. Ruler

5. A noiseless watch

6. Your school code

7. A valid photo identification and your Social Security Number

And, although you've heard it before, here's some last-minute advice: go to bed early the night before the exam and don't study all night—it won't do any good!

Good luck on your AP Physics 1 exam!

Solutions to Example Questions

a. (D) Adding a second bulb in series will increase the resistance of the circuit. Since there is only one loop for the current to travel, an increased resistance means a decreased current.

b. (D) Power can be calculated by $P = Fv$ or $P = \dfrac{\text{Work}}{\text{Time}}$. Either way you get the same answer. The force exerted to lift the 5 kg object at a constant speed must be equal to the object's weight, or approximately 50 N. When pulled at a constant speed of 3 m/s, the power is $P = (50\ N)(5\ m/s) = 250$ Watts. During the 3 seconds, since the object is traveling at 5 m/s, the object's displacement is 15 m. Therefore, the work done on the box is $W = (\text{Force})(\text{Displacement}) = (50\ N)(15m) = 750$ Nm; which makes the power equal to $P = \dfrac{\text{Work}}{\text{Time}} = \dfrac{750\ \text{Nm}}{3\ \text{seconds}}$.

c. (C) The acceleration of an object in free fall is always 9.8 m/s/s. It doesn't matter where in the trajectory the ball is or how fast, or at what angle, it was thrown. The acceleration is g, all day, every day.

d. (C) The time that the box will take to land depends only on the height of the table, which does not change. So the only thing that will affect the distance that the box travels before it lands is the speed with which it leaves the table. Since it is going three times as fast as the first box, it will land three times further from the edge of the table.

(continued)

e. (A) We are looking for the max speed which can be read right off the velocity vs. time graph. Choice (B) would work if you used the width of the block instead of the distance the block swings. (C) a motion detector doesn't give a time reading, and (D) the area under the position vs. time graph is not velocity.

f. (B) The equation for period of a simple pendulum is $T = T = 2\pi\sqrt{\dfrac{\ell}{g}}$.

A linear graph can be obtained if the students graph T^2 and ℓ. The slope of their graph would be $\dfrac{4\pi^2}{g}$.

g. (A) and (B) are correct. The equation for the resistance of a resistor is $R = \rho\dfrac{L}{A}$. You could increase the resistance of a resistor either by increasing the length of the resistor, decreasing the cross-sectional area or both. Be careful of the big pitfall of answers (C) and (D). If you are familiar with Ohm's law you might think that increasing the voltage or decreasing the current would also change the resistance, but it is not true! You can only change the resistance of the resistor by changing the physical dimensions or material of the resistor.

Key Terms

1. *Acceleration*: Acceleration tells us how much an object's velocity changes each second. When an object moves faster, its acceleration and velocity are in the same direction, and when an object slows down, the velocity and acceleration are in opposite directions. An object can experience acceleration moving at a constant speed. This happens when an object turns, as in circular motion. This is because velocity is a vector and has both magnitude and direction, and acceleration is defined as a change in either of these. Acceleration is measured in $\frac{m}{s^s}$. Think about acceleration being measured in $\frac{m/s}{s}$, which helps you remember that acceleration is how much velocity changes every second.

2. *Amplitude*: The amplitude is the distance from the midpoint of a wave to either the crest or the trough. For an oscillating system, the amplitude of the motion is the distance between the maximum displacement and the equilibrium position.

3. *Angular Momentum*: Angular momentum is an object's rotational inertia times its rotational velocity. For an object rotating a distance about a central axis of rotation, the change in an object's angular momentum is the product of its torque over a given time interval. Angular momentum is measured in $\frac{kgm^2}{s}$.

4. *Antinodes*: Positions on a standing wave with the maximum amplitude.

5. *Beats*: Interference that happens when two notes of unequal but close frequencies are played simultaneously. The beat frequency is the difference between the two frequencies.

6. ***Center of Mass:*** A unique point in an object or a system where one can consider all the mass as being concentrated. The center of mass can be used to analyze what will happen to the object or system under the influence of forces or torques.

7. ***Centripetal Acceleration:*** A "center-seeking" acceleration. An object moving in a circle experiences centripetal acceleration directed radially inward. The centripetal acceleration acts perpendicularly to the tangential velocity of the object and causes a change in direction rather than an increase in the object's speed.

8. ***Charge:*** A fundamental property of matter that is affected by an electric field. It is measured by an excess or deficit of electrons on an object. Charge (Q) can be positive or negative and is measured in coulombs (C).

9. ***Coefficient of Friction:*** A unitless quantity that tells us how "sticky" two surfaces are when rubbed past one another. This is a material-dependent quantity, and every pair of objects will have a unique coefficient of friction; the coefficient of static friction is larger than that of kinetic friction. The coefficient of friction is equal to the ratio of the frictional force to the normal force and is always less than one.

10. ***Coulomb's Law:*** Describes the force of attraction or repulsion between two electric charges. The electrostatic force is proportional to the magnitude of the product of the two charges and inversely proportional to the distance between them squared. The electric force is measured in Newtons (N) and is a vector quantity that has magnitude and direction. The direction of the electric force depends on the signs of the charges involved. For example, like charges repel and unlike charges attract.

11. ***Current:*** Conventional current is defined as the rate of flow of positive charge. Current is measured in Amperes (Amps). 1 Amp is defined as 1 coulomb of charge flowing per second.

12. ***Density:*** The mass per volume of a material, used as a measure of compactness of a substance. Density is measured in $\frac{kg}{m^3}$ and is represented by the Greek letter ρ (rho).

13. ***Displacement:*** A vector quantity, measured in meters, which tells us how far an object is from its initial position, and its direction. Displacement tells the change in position. Displacement is measured in meters (m).

14. *Distance*: A scalar quantity, measured in meters, telling how far an object has traveled.

15. *Doppler Effect*: The apparent change in a wave's frequency or wavelength because of the relative motion between the source of the wave and the observer.

16. *Elastic Collisions:* When two objects collide elastically, the momentum and the kinetic energy of the *system* are conserved. In an elastic collision, the two objects bounce off each other without any loss of kinetic energy.

17. *Electrostatic Force*: The force between objects caused by their electric charges. The electrostatic force is measured in Newtons (N) and is governed by Coulomb's law.

18. *Force*: Any push or pull (contact or not) from one object on another which could cause or contribute to the acceleration (or change in velocity) of one or both objects. Forces are a vector quantity and are measured in Newtons (N).

19. *Force of Gravity* (also called weight): The force of gravity is how hard the Earth pulls on you. The force of gravity depends on the mass of the Earth, your mass, and how far you are from the center of the Earth. For most situations the force of gravity on you is equal to $F_g = mg$, where g is the acceleration due to gravity $g = 9.8$ m/s^2.

20. *Free Body Diagram or Force Diagram*: A diagram showing the magnitude and the direction of all forces acting *on* an object.

21. *Free fall:* An object's movement under the influence of only the force of gravity, where the acceleration is due to gravity (g), where $g = 9.8$ m/s^2. Examples of free fall would include an object traveling (for a moment) upward.

22. *Frequency*: The inverse of period, the number of cycles or of wavelengths passing a position every second. Frequency is measured in $\frac{1}{s}$ or Hertz (Hz).

23. *Gravitational Field*: A model explaining how massive bodies exert gravitational forces on other bodies. A gravitational field is defined as the gravitational force created by a massive central object divided by the mass of any other object at a given distance from the center of

the object. On the Earth's surface, the gravitational field strength is 9.8 N/kg.

24. *Gravitational Force*: The universal force existing between two objects because of their masses. It is directly proportional to the product of the masses and is inversely proportional to the square of the distance between the objects (measured center to center). Gravitational force acts equally and oppositely on the two objects involved in the inter-action and, unlike the electrostatic force, is always an attractive force. This force is measured in Newtons (N).

25. *Gravitational Potential Energy*: Energy related to the position of an object above or below a reference point. Gravitational potential energy cannot be possessed by a single object; rather it is energy stored between two objects due to their relative positions and masses.

26. *Impulse*: The change in an object's momentum which is equal to the product of the net force on an object and the time interval over which the force was applied. Impulse is measured in kg · m/s (like momentum) or in N · s; these are equivalent units. The direction of the impulse will be the same as the direction of the net force on the object.

27. *Inelastic Collision*: When two objects collide inelastically, the momen-tum of the *system* is conserved, but kinetic energy from the system will be transferred to other non-mechanical forms (such as heat, sound, or deformation of the objects). A perfectly inelastic collision is when objects stick together and do not bounce. If unsure about the kind of collision, compare the kinetic energy before and after the col-lision. If kinetic energy is lost, the collision is inelastic.

28. *Inertia*: The property of an object's mass that causes it to move in a straight line at a constant velocity. It may also be thought of as the object's resistance to acceleration as used in Newton's Second Law $\left(a = \dfrac{F}{m} \right)$. This property of mass is different from the gravitational property.

29. *Interference*: When two waves cross over and through each other, they will interact with one another. This interaction is called interfer-ence. Constructive interference is when the crests of two waves meet up, and they add to form (for an instant) a larger wave. Destructive interference is when a crest and a trough from two waves meet up and add to form (for an instant) a much smaller wave.

30. *Internal Energy*: This can have two meanings depending on the situation. First, microscopic internal energy is related to the temperature of the object. For example, a box sliding across the floor comes to a stop. All of the box's initial kinetic energy was turned into internal energy (heating up the box and the floor). Second, internal energy could be referencing the total stored potential energy of the system.

31. *Kinetic Energy*: The energy associated with the motion of an object. Kinetic energy is measured in Joules or $kg \cdot \dfrac{m^2}{s^2}$. There are two types of kinetic energy: *rotational kinetic energy,* which is associated with rotating objects, and *translational kinetic energy*, which is associated with objects experiencing a translational motion.

32. *Kinetic Frictional Force:* A resistive force that opposes the sliding motion of an object. This force exists between the surface and the sliding object when the object is *moving* relative to the surface. The kinetic frictional force is parallel to the surface and opposite to the direction of motion.

33. *Kirchhoff's Laws*: Kirchhoff's laws are used to describe current flow and potential difference throughout DC circuits. They simplify to conservation laws, specifically conservation of energy and conservation of charge.

34. *Lever Arm* (also called the moment arm): A lever arm is a quantity used in rotating systems and is defined as the distance between the rotation point and where a given force is applied.

35. *Longitudinal Wave*: A longitudinal wave carries energy by vibrating the particles of the medium parallel to the direction that the energy is being carried through the medium.

36. *Mass*: The amount of matter in an object. Mass is measured in kilograms (kgs) and will be the same no matter where you travel in the universe. Gravitational mass measures the amount of force an object will feel inside a gravitational field and inertial mass measures an object's resistance to being accelerated by a force. The measure of the two masses is equivalent as far as we can currently measure.

37. *Mechanical Energy:* The sum of the kinetic and potential energies of a system. Mechanical energy deals with the energy associated with an object's motion and relative position within a system.

38. ***Momentum***: An object's momentum is equal to the product of its mass and velocity. Momentum is a vector quantity, so be *very* careful to note the direction of momentum. It is always the same direction as the object's velocity. Momentum is measured in kg · m/s.

39. ***Newton's First Law of Motion:*** Also called the Law of Inertia, Newton's First Law of Motion states that an object at rest will remain at rest, and an object in motion will remain in motion, at a constant velocity, unless acted on by an outside force.

40. ***Newton's Second Law of Motion***: This law states that the net force (or the total force) on an object will cause an acceleration that is proportional to the force and inversely proportional to the mass of the object. This acceleration is always in the direction of the net force. For any system of constant mass, a larger net force will produce a larger acceleration. With a given net force, the resulting acceleration will be less for a more massive object.

41. ***Newton's Third Law of Motion***: This law states that every action has an equal and opposite reaction. In other words, the force that object A exerts on object B is the same size and in the opposite direction to the force that object B exerts on object A. Forces described by Newton's Third Law are known as action-reaction pairs. Action-reaction pairs act on different objects and NEVER cancel out. The easiest way to identify an action-reaction pair is to remember that you are looking at two objects only.

42. ***Nodes***: The stationary point(s) on a standing wave where the amplitude is a minimum.

43. ***Normal Force***: A contact force existing between two objects that keeps one object from invading another object's space. *Normal* means perpendicular, and the normal force, by extension, is a force perpendicular to a surface. If a box sits on a table, the force from the table can be called F_{Table}, or since the table will push up on the box (perpendicular to the surface of the table), it is also a normal force.

44. ***Object:*** A physical body or collection of matter. An object could be a single electron or a box or an elephant depending on the problem and the situation being analyzed.

45. ***Parallel***: Circuit elements are connected in parallel if the path for the current splits and then comes back together. Elements connected in parallel will have the same potential difference across them.

46. *Period*: The time for a complete cycle or the time for a full wavelength to pass a position. Period is measured in seconds (s). The period of an oscillating spring system is dependent upon the mass of the object and the spring constant. The period of an oscillating pendulum is dependent on the pendulum's length and the gravitational field's strength in which the pendulum exists.

47. *Position*: The location of an object is relative to a chosen or given coordinate system. Position is measured in meters (m).

48. *Power*: The rate at which work is done, or energy is used per second. This can be simplified to the product of force and velocity. For electrical power, this can be simplified to the product of the current and the potential difference. Power is measured in Joules/seconds or watts.

49. *Projectile motion*: An object in free fall that also has a horizontal component to its motion, so that it travels horizontally with a constant velocity while undergoing an acceleration (equivalent to free-fall acceleration or g [9.8 m/s² on the surface of the Earth]) at the same time.

50. *Resistance*: The opposition to the flow of electric charge through a circuit element. Resistance is measured in ohms (Ω). Resistance can be affected by the material (resistivity), the temperature, the length of an object, and the cross-sectional surface area through which the charge flows.

51. *Resistivity*: The property of a material that tells us what the resistance would be of a cubic meter of that material. Resistivity is based on the physical dimensions and material of an object. Resistivity is measured in ohm meters ($\Omega \cdot$ m).

52. *Restoring force*: A restoring force is any force that pushes an object back toward an equilibrium position.

53. *Rotational Acceleration*: Analogous to linear acceleration, rotational acceleration is the rate of change of angular velocity. It is measured in radians/s².

54. *Rotational Displacement*: Analogous to linear displacement, rotational displacement is how many radians an object has rotated through as it rotates. Rotational displacement is measured in radians (θ).

55. *Rotational Inertia*: Also called the moment of inertia. It is a scalar that is dependent on the mass of the object and how that mass is arranged. Rotational inertia is a measure of an object's resistance to a

change in its rotation around a given pivot point. Rotational inertia is measured in kg · m^2.

56. **Rotational Kinetic Energy**: The rotational analogue to translational kinetic energy. The energy associated with the rotational motion of an object. Rotational kinetic energy is proportional to an object's moment of inertia times the square of the angular velocity. Rotational kinetic energy is measured in Joules.

57. **Rotational Velocity**: Analogous to linear velocity, rotational velocity is the rate of change of angular position. Rotational velocity is measured in radians/s.

58. **Scalar**: A number without direction. This quantity has magnitude only.

59. **Series**: Circuit elements connected in series form a single path for the current to flow through.

60. **Speed**: The distance traveled per unit of time. Speed is a scalar quantity and is a measurement of how fast an object is going but does not indicate direction. Speed is measured in m/s.

61. **Spring Constant**: The spring constant, k, tells us how stiff or tough a spring is; the larger the k value for a spring, the more difficult it is to stretch that spring. The spring constant is measured in N/m.

62. **Spring Force**: The force exerted on an object by a compressed or stretched spring. Spring force is a restoring force, which means it is oriented toward equilibrium. A Hooke's law spring is a special spring where the restoring force is proportional to the distance the spring is stretched or squished.

63. **Spring Potential Energy**: Also known as elastic potential energy. This energy is equivalent to the work done to deform the elastic object, such as a spring, and occurs when it is stretched or compressed and is related to the spring's relative displacement. Just as with gravitational potential energy, spring potential energy is not stored in a single object. Spring potential energy is stored between the object and the spring.

64. **Standing Waves**: A wave that appears to stay in one place on a string instead of moving up and down the string. Standing waves are formed from interference.

65. *Static Frictional Force:* A resistive force that opposes the sliding motion of an object. This force exists between the surface and the sliding object when the object is *at rest* relative to the surface. The static frictional force is parallel to the surface and opposite to the direction of motion, *if motion relative to the surface were to occur.*

66. *Superposition*: The principle of superposition allows us to add and subtract wave forms that constructively or destructively interfere with one another.

67. *System*: A group of objects that can be treated as a single object. This may seem foreign now, but if you think about it, we're really doing this all the time. Imagine a simple ball; although it looks like one single object, it is really made up of molecules of atoms and of protons, electrons, and neutrons—all of which is treated as one single ball. Before one starts a problem, it is important to choose and identify the system being worked with.

68. *Tension*: A pulling force exerted by a string or a rope. This force is transmitted through the string when the string is pulled at both ends. The force is directed along the string and pulls equally on the objects on either end of the string. Tension is uniform throughout the string or rope.

69. *Torque*: The rotational analog to force. Torque tells us how much rotational acceleration will be caused when a certain force is applied at a given distance from the axis of rotation (also known as the lever arm or moment arm). Torque is measured in N · m but should not be equivocated with Joules.

70. *Translational Kinetic Energy:* The energy that is associated with translational motion, which occurs when an object's center of mass moves. Translational kinetic energy is proportional to an object's mass times the square of the velocity.

71. *Transverse Wave*: A wave that carries energy by vibrating the particles of the medium perpendicular to the direction that the energy is being carried through the medium.

72. *Universal Gravitational Constant*: G is equal to $6.67 \times 10^{-11} \dfrac{N \cdot m^2}{kg^2}$.

73. *Vector*: A quantity that has magnitude and direction.

74. **Velocity**: The rate of change of displacement. Velocity is a vector that has a magnitude and a direction. Velocity tells how fast something moves and the direction in which it moves. Velocity is measured in m/s.

75. **Voltage**: A measurement of the electrical potential energy per coulomb of charge. Voltage is related to the amount of work done to move a charge through an electric field. Voltage is measured in volts (V).

76. **Wave:** A disturbance in a medium. The disturbance carries energy through the medium without permanently disturbing the matter in the medium.

77. **Wavelength**: The measured distance from peak to peak, or trough to trough, on a wave. (The wavelength can also be measured between any successive identical points on a wave.) Wavelength is measured in meters.

78. **Weight**: The force of gravity on an object. This is different than mass. The weight is how much the Earth (or whatever planet is closest) pulls on that mass. Your weight could be significantly different depending on where you are in the universe. Weight is measured in Newtons (N).

79. **Work**: The change in energy of a system. Work is also equal to the net force acting on an object times the distance over which that force is applied. (The distance is measured as that being parallel to the force.) Work is measured in Joules or Newton · Meters (N · m).

80. **Work Energy Theorem**: When a net force works on an object, it causes the kinetic energy of the object to change. The amount of work done on the object is equivalent to the change in the object's kinetic energy.

The Big Ideas

I. INTRODUCTION

The AP Physics 1 curriculum is designed to test your understanding of what the College Board labels the "Big Ideas." The Big Ideas are overarching topics and themes that run through the physics curriculum. The College Board has moved away from calculation-based questions and more toward questions requiring more explanation. In addition, the number of questions on the exam has decreased, allowing more time for contemplation, explanation, and writing.

II. THE BIG IDEAS

A. Big Idea 1: Objects and systems have properties such as mass and charge. Systems may have internal structure.[1]

Big Idea 1 focuses on objects and systems. When solving a problem in AP Physics 1, you will need to understand that everything is made up of smaller things. In a problem about a box, you will usually treat the box as an object, although it is really a system of molecules made up of atoms, which are made up of protons, neutrons, and electrons, which in turn are made up of even smaller things. You will need to decide when to use a system solution and when objects can be used to solve problems.

Big Idea 1 is also about charge. Charge is conserved and comes in fundamental packets. The smallest observable charge is the charge of an electron, which is also called the elementary charge, where $e = 1.6 \times 10^{-19}$ C. There are two kinds of charge: positive

---

[1] From the College Board AP Physics 1 Framework, published 2014.

and negative. Neutral objects have an equal amount of both or, in special cases, zero total charge.

Big Idea 1 notes the two properties of mass: gravitational attraction and inertia. Although the two properties are different, it can be proven experimentally that the object's gravitational mass and its inertial mass are equivalent.

In addition, Big Idea 1 discusses how various materials have different properties associated with their molecular make-up such as density and resistivity.

B. Big Idea 2: Fields existing in space can be used to explain interactions.[2]

Big Idea 2 is a focus in AP Physics 2; however, gravitational fields are covered in AP Physics 1. Fields represent "action at a distance" forces where direct contact is not required. An understanding of vector fields is needed, specifically the gravitational field that is created by any object which has mass. This field radiates spherically outward from an object and represents the gravitational force on a second object divided by the second object's mass. The strength of the gravitational field is related to the position within the field and is equal to the acceleration of gravity (g) at that point.

C. Big Idea 3: The interactions of an object with other objects can be described by forces.[3]

Big Idea 3 covers material in AP Physics 1. You must have an understanding of kinematics (motion) and what causes motion (forces). Vectors appear here because quantities discussed in motion and forces are vector quantities. You must be able to interpret and draw free body diagrams and to determine when forces are applied on an object. Objects cannot exert forces on themselves, and all interactions involve two objects. It is important to understand situations involving static and dynamic equilibrium and to know that even objects at rest can have forces on them. It is important to know the difference between acceleration and velocity and to understand they are related, but different. Newton's laws are critical knowledge: how inertia works, that accel-

---

[2,3] From the College Board AP Physics 1 Framework, published 2014.

eration is directly proportional to the net force and that every force has an equal and opposite force.

Big Idea 3 also discusses spring forces and the gravitational force acting as restorative forces causing oscillations for springs and pendulums. This links the concepts of oscillatory motion and forces together.

In addition to contact forces such as tension, friction, normal, and spring forces, where two objects touch one another to exert the force, Big Idea 3 discusses action-at-a distance forces like the gravitational force and the electrostatic force. The ability to discuss the similarities and differences between the electrostatic and the gravitational forces is important—noting specifically both follow an inverse square law and are directly proportional to each of the fundamental quantities involved (either two objects that have mass as in the case of gravitation or two objects that have charge as in the case of electrostatic forces).

Forces can change the energy of a system by doing work on that system. This can involve a change in translational kinetic energy involving a force parallel to the displacement it causes, or it can involve a change in rotational kinetic energy where the force is perpendicular to the lever arm and the resulting torque causes rotation. In both translational and rotational systems, it is important to note how forces relate to changes in momentum. In translational motion, the direction of the net force on an object will be the same as the direction of the change in momentum. Forces can also cause a system to rotate by applying torque to that object. Applying a force that causes a torque will change the angular momentum of the object in the same way that applying a force to an object can change the object's linear momentum.

D. Big Idea 4: Interactions between systems can result in changes in those systems.[4]

Big Idea 4 focuses on changes in systems. Knowing that linear motion can be described by displacement, velocity, and acceleration of the center of mass is important. Acceleration is equal to the change in velocity as a function of time, and velocity is the rate of change of position as a function of time. You must be able

[4] From the College Board AP Physics 1 Framework, published 2014.

to explain how acceleration is caused by a net force, and why it is in the same direction as the net force.

Big Idea 4, reminds us that change in linear momentum for an object is the mass of the object times the change in velocity of that object. Change in linear momentum is also equal to the product of the force on the object and the time over which the force acts. Likewise, a change in angular momentum is caused by a torque, and is defined as the product of the average net torque on an object and the time over which that torque is acting. Forces on an object or a system can change the energy of the system just as torques can change the rotational kinetic energy.

E. **Big Idea 5: Changes that occur as a result of interactions are constrained by conservation laws.**[5]

Big Idea 5 focuses on conservation laws. The ability to define an open and a closed system, and apply conservation laws to these systems is important. The quantities of energy, momentum, mass, and charge are conserved in an isolated system. In an open system, one or more of these quantities are exchanged into and out of the system. Interactions can be either forces or just a transfer of some quantity (i.e., charge being added to the system from an outside source).

Deciding what is part of the system may simplify your analysis. Problems will involve the conservation of mechanical energy, which includes kinetic and potential energy. Work is the transfer of energy and is done by external forces acting on an object or system through some distance. The rate energy is transferred into or out of a system is power.

Big Idea 5 covers the conservation of both linear and angular momentum. In a linear system, momentum is conserved in all types of collision (elastic, inelastic, and perfectly inelastic) while kinetic energy is conserved only in elastic collisions. The momentum of an object will not change if the net force acting on the object is zero, and internal forces will not change the momentum of an object or system. Likewise, if the net torque is zero acting on a rotating object or system, there will be no change in the object's angular momentum.

---

[5] From the College Board AP Physics 1 Framework, published 2014.

The rules governing the conservation of energy in DC electrical circuits are known as Kirchhoff's rules. Since electric charge is conserved, electric current must be conserved at each junction in a circuit.

F. **Big Idea 6: Waves can transfer energy and momentum from one location to another without the permanent transfer of mass and serve as a mathematical model for the description of other phenomena.**[6]

Big Idea 6 focuses on wave phenomena. There are two different types of waves: longitudinal and transverse. Mechanical waves, such as sound waves, require a medium to propagate while electromagnetic waves, which travel as fields, do not require a medium and can travel through the vacuum of space. Waves transfer energy without permanently displacing the medium. Periodic waves have an amplitude, frequency, wavelength, speed, and energy. The wave equation relates the speed of a wave as it propagates to its frequency and wavelength as the product of these two quantities. The ability to interpret and create these equations for different waves is important. You must be able to explain that wave pulses interfere with each other via superposition and standing waves are the result of a series of wave pulses interfering with each other. The wavelengths of standing waves are determined by the length of the string or the length of the tube containing the wave.

G. **Big Idea 7: The mathematics of probability can be used to describe the behavior of complex systems and to interpret the behavior of quantum mechanical systems.**[7]

Big Idea 7 is discussed exclusively in AP Physics 2.

[6,7] From the College Board AP Physics 1 Framework, published 2014.

PART II

CONTENT REVIEW

Kinematics

MOTION

1. All motion is relative. This means that how you measure someone else's motion depends on where you are standing and your own motion, compared to something or someone else.

   EXAMPLE: Imagine an elephant riding on a circus train traveling east at 5 miles per hour. If you are standing on the ground watching the train go by, you would say that the elephant was traveling east at 5 miles per hour. However, if you were standing on the train *with* the elephant, you would say that relative to you, the elephant was at rest.

2. In order to discuss motion, we need to have chosen a frame of reference. A frame of reference includes a coordinate system that can be used as a reference to describe an object's motion; it is in reference to this coordinate system that values of position and time are measured. This can be an arbitrary set for each given situation. To start any problem with motion, the first step is always to choose a point to call zero, from which you will measure all motion.

3. This zero point is often called the *origin*. You have to be careful, however, both in your mind and in your answer because there is a difference between the spot you call the origin (0, 0) and the origin, or starting point, of the object's motion.

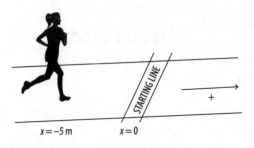

x = −5 m   x = 0

EXAMPLE: Consider a runner getting ready for a race. Let's choose the origin, to be the starting line. That's where we'll put our zero position. However, this is a big race, with lots of runners, so when the starting gun is fired, our runner is 5 meters behind the starting line. So the origin, or starting point, of her motion is $x = -5m$.

4. Once you choose your origin, decide which direction you will label as the positive direction: up or down for vertical motion, and left or right for horizontal motion. It doesn't matter which you choose, but remember to remain consistent with that choice throughout the solution to the problem.

5. As we begin to investigate motion, we will be measuring both *vectors* and *scalars*.

   a. A **scalar** is a quantity that has magnitude only. (A number that represents "how much" of something there is.) Some examples of scalars include:

      i.   Currency: 5 dollars

      ii. Temperature: 25 Kelvin

      iii. Objects/things: 15 people

      iv. Mass: 20 kg

      v.   Speed: 5 m/s

   b. A **vector** is a quantity that has both magnitude and direction. (A number and a direction. "How much" and "which way" is a more complete description of motion). Vectors can always be represented by arrows that show direction and relative magnitude depending on the length of the arrow. Some examples of vectors include:

i.   Displacement: 20 meters *west*

ii. Velocity: −5 m/s

6. An observer in a particular reference frame can describe the motion of an object using such quantities as position, distance, displacement, velocity, speed, and acceleration.

   a. *Position* is measured relative to the origin. It can be positive or negative and is represented by the letter x or y on a horizontal or vertical number line.

   b. *Distance* traveled is a scalar and represents how much ground an object has covered or how far it has gone during the time interval analyzed.

   c. *Displacement* is a vector and represents the space between where the object started and where it ended. In other words, the object's overall change in position.

   i.   If Carl walks 4 steps forward and then 3 steps back, the total distance he has traveled is 7 steps. (Remember distance is a scalar, so direction is not important.) However, compared to his starting point, Carl is only 1 step in front of where he started, so his total displacement is 1 step forward. (For displacement, a direction is necessary.)

   ii. If Carl decides to turn a corner, the difference between distance and displacement becomes a little more complex. Carl now walks 4 steps east and then 3 steps north. The total distance that Carl covered is still 7 steps. To find Carl's displacement, we need to find out how far Carl is from where he started.

3 Steps North

4 Steps East

Note that because he made a 90° turn, his change in position can be found by drawing an arrow from his starting point to his ending point to represent the total displacement.

iii. When drawing arrows to represent vectors, pay attention to both the direction and magnitude of the vector. The arrow points in the direction of the vector, while the length of the arrow indicates the relative magnitude. When finding a resultant, or total vector, as a result of two vectors added together (the components), one must always orient the component vectors "Tip to Tail." This means you should never have two tails of component vectors touching nor should you have two tips touching. The resultant vector is drawn from beginning point to end point.

iv. In this example, vectors are represented as arrows that form a right triangle with the *x* and *y* displacement vectors represented in their respective directions of 4 steps east and 3 steps north and with the total displacement vector represented by the hypotenuse of the triangle. Because this forms a right triangle, we can use the Pythagorean theorem to calculate the magnitude of his displacement.

$$\text{Displacement} = \sqrt{X^2_{\text{Component}} + Y^2_{\text{Component}}}$$

In the case of Carl turning the corner, although he walks 4 steps east and then 3 steps north, his displacement is $\sqrt{4^2 + 3^2}$ = 5 steps.

v. Because displacement is a vector, it always needs a direction. Trigonometry is used to calculate a direction for the vector.

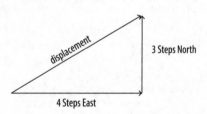

When calculating the displacement vector, the total displacement vector should point from the original position (the bottom left of the triangle) up toward the final position (the upper right of the triangle).

vi. To describe this displacement with respect to its starting point, find the angle of the resultant vector (the hypotenuse) with respect to its first motion, the horizontal line as shown in the diagram. Remember the tangent of an angle is equal to the opposite side divided by the adjacent side.

$$\tan\theta = \frac{\text{Opposite}}{\text{Adjacent}}$$

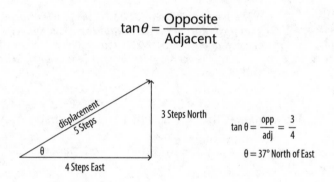

7. *Speed* is how fast an object is moving. It is a scalar quantity measured in meters per second (*m/s*).

   a. *Average speed* is defined as the total distance divided by the change in time.

   b. $\text{Speed} = \dfrac{\text{total distance}}{\Delta t}$

   c. *Instantaneous speed* is how fast an object is traveling at a certain instant. Although your *average* speed on a long trip might be 40 miles per hour, at any one point along the trip, your *instantaneous* speed might have been 70 miles per hour.

8. *Velocity* is speed with a direction. It is a vector quantity defined as the rate of change of position. It is usually measured in meters per second (m/s) and is represented by the letter *v*.

   a. Average Velocity $= \dfrac{\text{Displacement}}{\Delta \text{ time}} = \dfrac{\text{change in position}}{\Delta \text{ time}}$

   The Greek symbol Δ (delta) represents a change in quantity. In this case, it represents the change in time (Δt), but later it will represent the changes in other vector quantities. You can calculate a change in any vector quantity by subtracting the initial value from the final value: $\Delta t = t_f - t_i$.

Test Tip

*The symbol **v** is used for both speed AND velocity. It is important to understand that there is a difference between speed and velocity. When you are dealing with velocity, the direction is very important and cannot be ignored.*

9. *Acceleration* is the rate at which velocity changes. Acceleration is a vector quantity measured in meters per second $\left(\dfrac{\frac{m}{s}}{s} \right)$.

   a. Units of acceleration are often written as $\dfrac{m}{s^2}$ which can be confusing. What is a squared second? A better way to think about acceleration is to think of the units as $\dfrac{\frac{m}{s}}{s}$. This will help you to remember that acceleration is how many m/s the velocity changes for every second that passes. Acceleration is represented by the letter *a*.

10. There are three major motion equations that you'll need to solve, understand, and discuss. These equations can be used for objects experiencing a constant acceleration, and this is an important distinction. Here is an example to help you understand:

    a. You are on a road trip with your friends. The car is cruising at a rate of 20 mph for 20 miles. The driver then decides to accelerate, and the car cruises along at a rate of 40 mph for another 20 miles. What is the average velocity?

You cannot use the equations of motion for this problem because the car did not undergo a constant acceleration. Rather, the car went one speed for a given amount of time and then went another speed for another amount of time.

The problem is still solvable by understanding that an average velocity is equal to a total displacement over a total time interval.

The car travels a total distance of 40 miles, but the total time needs to be determined.

The time it would take to travel the first 20 miles would be equal to $\Delta t = \dfrac{\text{displacement}}{\text{average velocity}} = \dfrac{20 \text{ miles}}{20 \text{ mph}} = 1 \text{ hour}$. The time it would take to travel the second 20 miles would be equal to $\Delta t = \dfrac{\text{displacement}}{\text{average velocity}} = \dfrac{20 \text{ miles}}{40 \text{ mph}} = .5 \text{ hours}$. The total time therefore would be 1.5 hours, and the average velocity would be $\text{Average Velocity} = \dfrac{\text{Displacement}}{\Delta \text{ time}} = \dfrac{40 \text{ miles}}{1.5 \text{ hours}} = 26.7 \text{ mph}$.

If the car had gradually sped up from rest with a constant acceleration to some final velocity, then the equations of motion would be applicable.

11. The three equations of motion for objects undergoing a constant acceleration are:

$$x_f = x_0 + v_0 t + \frac{1}{2} a t^2$$
$$v = v_0 + at$$
$$v_f^2 = v_0^2 + 2a(x_f - x_0)$$

[In each of these equations, x represents position, v is velocity, a is acceleration, and t is time. The subscripts $_0$ and $_f$ represent the initial and final values of position and velocity.]

Test Tip

Every equation requires an initial velocity. If it is not explicitly stated, assume it is zero.

When solving a numerical problem on the AP Physics 1 exam, it will be helpful to make a list of the quantities given and then compare that list to what you are asked to calculate.

B. GRAPHS

1. The equations above are derived from the use of graphs to represent an object's motion. Graphs are valuable tools, rich with information about an object's motion. You will need to be able to read, interpret, and create graphs of one-dimensional motion.

2. The position versus time graph is probably the most intuitive. It maps where the object is located at each given moment in time.

Position vs. Time Graph	Description of Motion
❶ *x*(m) 6 ⊢———→ *t*(s)	Charlie sits in his lawn chair 6 m from the origin for time (*t*). Note the graph shows that at each instant, Charlie's position was always a positive 6 m away from the origin. This produces a graph with a positive *y*-intercept and a flat horizontal line. Note also the slope is zero, indicating no change in position with respect to time.

Position vs. Time Graph	Description of Motion
❷ 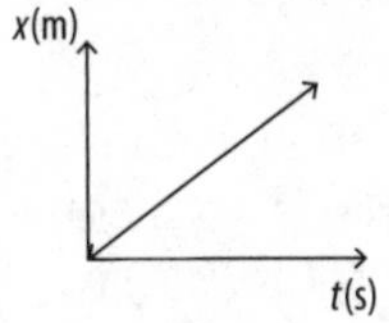 *x*(m) *t*(s)	Charlie starts at the origin and moves away in a positive direction at a constant speed for time (*t*). Here, the position vs. time graph shows Charlie starts with a position of zero when the clock starts, but his position is farther and farther away from the origin as time goes on. This produces a graph that has a positive constant slope. Note the slope is constant and positive, indicating a constant change in position with respect to time, in the positive direction.
❸ 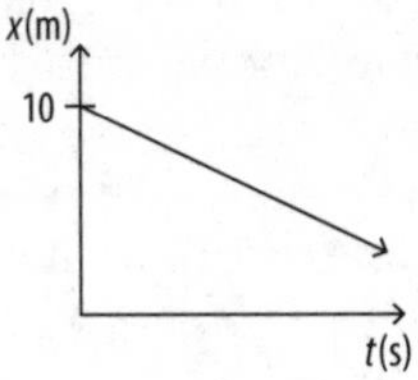 *x*(m) 10 *t*(s)	Charlie starts 10m from the origin and moves back toward the origin at a constant speed for time (*t*). Note the intercept on the graph tells us the starting position at time equals zero, and since the *y*-value gets smaller, it means that Charlie is getting closer to the origin as times goes on. This produces a graph with a constant negative slope, showing Charlie is moving in the negative direction (or in the opposite direction compared to the positive direction on our coordinate system). Note the slope is constant and negative, indicating a constant change in position with respect to time in the negative direction.

Position vs. Time Graph	Description of Motion
4	Charlie starts at the origin at rest and accelerates in the positive direction for time (*t*).
	Note the intercept on the graph is the origin, and since Charlie is accelerating away, he will be getting further and further from the origin at an increasing rate.
	Note the slope starts as zero, becomes slightly positive, and increasingly more positive. This slope is not constant, resulting in a curved rather than a linear function. The increasing steepness of the slope indicates the object is speeding up (the relative change in position with respect to time gets higher and higher), and the slope is positive, showing this is happening in positive direction.
5	Charlie starts 10m from the origin at rest and accelerates with a negative acceleration for time (*t*).
	Note the intercept on the graph tells us the starting position at time equals zero, and since he is accelerating in the negative direction, he will be getting closer and closer to the origin, at an increasing rate.
	Note the slope starts as zero, becomes slightly negative, and increasingly more negative. This slope is not constant, resulting in a curved rather than a linear function. The increasing steepness of the slope indicates that the object is speeding up (the relative change in position with respect to time gets higher and higher), and the slope is negative, showing this is happening in the negative direction.

Position vs. Time Graph	Description of Motion
❻ x(m) t(s)	Charlie starts at the origin with a positive initial velocity and accelerates in the negative direction for time (t). Notice because Charlie's velocity and acceleration are in opposite directions, he ends up turning around and coming back toward where he started. Note the slope starts as positive, becomes less and less steep until it reaches a point where it is zero. The positive slope indicates the object is moving forward, and the decreasing steepness tells us he is slowing down. The point where the slope is zero is the turn-around point; this is where Charlie momentarily stops as he reverses his direction of travel. The slope then becomes negative and becomes increasingly steeper. The now negative slope tells us he is now moving in the negative direction, and the increasingly steep slope indicates he is speeding up in the negative direction.

Position vs. Time Graph	Description of Motion
❼ x(m) 10	Charlie starts at 10m from the origin with a negative velocity and a positive acceleration for time (t). Notice again as the result of the velocity and accelerations being in opposite directions, Charlie ends up turning around and heading back toward where he came from. Note the slope starts as negative, becomes less steep until it reaches a point where it is zero. The negative slope indicates the object is moving backward, and the decreasing steepness tells us he is slowing down. The point where the slope is zero is the turn-around point; this is where Charlie momentarily stops as he reverses his direction of travel. The slope then becomes positive and becomes increasingly steeper. The now positive slope indicates he is now moving in the positive direction, and the increasingly steep slope indicates he is speeding up in the positive direction.

C. **SLOPE OF THE GRAPH**

1. In addition to the general shape of a graph, it is important to understand the *slope* of the graph. The slope of the graph tells some very important information about an object's motion.

Memorizing specific graphs will not help; however, understanding the slope of a graph can help you analyze and interpret motion for objects even on graphs you have never seen before. Thus, learning to understand graphs will be a more useful way to spend your study time.

2. Remember on the position vs. time graph, we noticed the slope (calculated as rise/run = slope = $\tan\theta = \dfrac{y_2 - y_1}{x_2 - x_1}$) is measured in units of meters per second (m/s). This means the slope of the position vs. time graph can tell us about the velocity (rate or speed as well as direction) of the motion.

3. This knowledge helps us to expand the chart above to include not only position vs. time graphs, but also velocity vs. time graphs.

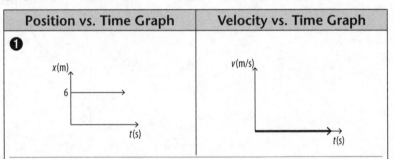

Position vs. Time Graph	Velocity vs. Time Graph

Description of Motion

Charlie sits in his lawn chair 6 m from the origin for time (t).

Understanding the position vs. time graph has a slope of zero, we will recognize that Charlie must have had a zero velocity. Because Charlie's position remains constant, this means he was not moving. Thus, the velocity vs. time graph shows his velocity was zero at each moment in time.

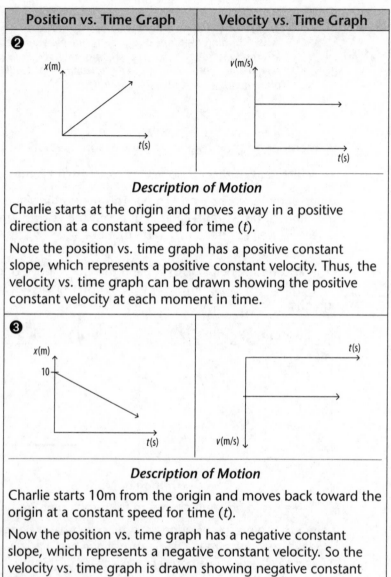

Position vs. Time Graph	Velocity vs. Time Graph

❷

x(m)

t(s)

v(m/s)

t(s)

Description of Motion

Charlie starts at the origin and moves away in a positive direction at a constant speed for time (*t*).

Note the position vs. time graph has a positive constant slope, which represents a positive constant velocity. Thus, the velocity vs. time graph can be drawn showing the positive constant velocity at each moment in time.

❸

x(m)

10

t(s)

t(s)

v(m/s)

Description of Motion

Charlie starts 10m from the origin and moves back toward the origin at a constant speed for time (*t*).

Now the position vs. time graph has a negative constant slope, which represents a negative constant velocity. So the velocity vs. time graph is drawn showing negative constant velocity at each moment in time.

Position vs. Time Graph	Velocity vs. Time Graph

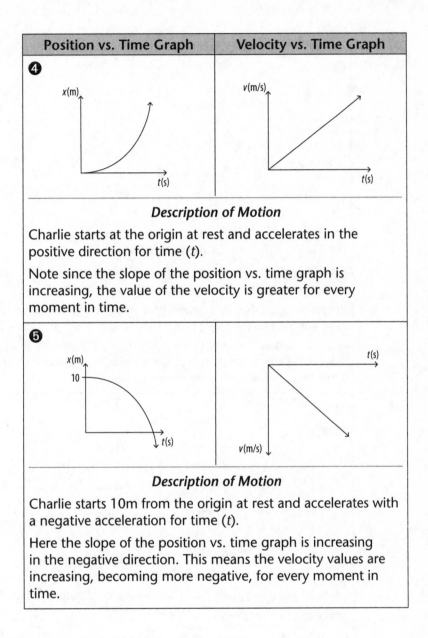

4

Description of Motion

Charlie starts at the origin at rest and accelerates in the positive direction for time (*t*).

Note since the slope of the position vs. time graph is increasing, the value of the velocity is greater for every moment in time.

5

Description of Motion

Charlie starts 10m from the origin at rest and accelerates with a negative acceleration for time (*t*).

Here the slope of the position vs. time graph is increasing in the negative direction. This means the velocity values are increasing, becoming more negative, for every moment in time.

Position vs. Time Graph	Velocity vs. Time Graph
❻ 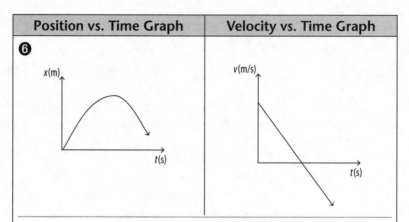	

Description of Motion

Charlie starts at the origin with a positive initial velocity and accelerates in the negative direction for time (t).

In the position vs. time graph, the slope starts out positive, gets less and less steep, at one moment is zero, and then the slope becomes more and more negative. This means that the velocity values start with a positive velocity, decrease through zero, and then become increasingly negative.

❼ 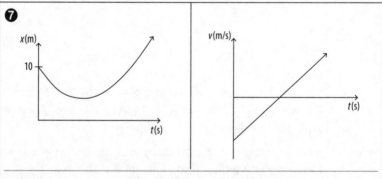	

Description of Motion

Charlie starts at 10m from the origin with a negative velocity and a positive acceleration for time (t).

In the position vs. time graph, the slope is originally negative, then the slope gets closer to zero, becomes zero at an instant, and then the slope becomes more positive. This means the velocity values start negative, get closer and closer to zero, pass through zero, and then become more positive.

4. Notice the slope of the position vs. time graph shows the velocity at any point in time.

5. The two graphs in No. 6 are particularly interesting. Notice although the object is initially moving away from the origin (in the positive direction), the slope becomes flatter (but still positive) meaning that the object is slowing down. The slope becomes completely flat (zero slope) meaning that the object stops; it then becomes increasingly negative, meaning that the object speeds back up in the negative direction.

6. In the same way the slope of the position vs. time graph is significant, the slope of the velocity vs. time graph is also significant. The slope of the velocity vs. time graph (slope = $\dfrac{v_2 - v_1}{t_2 - t_1} = \dfrac{\frac{m}{s}}{s}$) means the slope of the velocity graph with units of m/s/s or m/s^2 is equal to the acceleration of the object.

D. AREA UNDER THE GRAPH

1. Another feature of graphs that is useful in describing an object's motion is the area enclosed by the line on the graph and the axis. In the case of constant velocity (a flat line on the velocity vs. time graph), the area is in the shape of a rectangle, whose area can be found by multiplying the length (the value on the x-axis) by the height (the value on the y-axis). Note that the units of this product are meters/second × seconds = meters.

2. Thus, the area under the velocity vs. time graph is equal to the displacement or the change in position during that time interval.

3. For an object moving at a constant speed in a positive direction, the area is equal to the change in position $\Delta x = v \times t$.

4. For an object starting from rest and moving with a positive acceleration, the area under the velocity vs. time graph is equal to the area under the triangle (1/2 base × height), $\Delta x = \dfrac{1}{2} v_f t_1$.

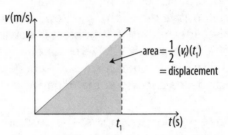

E. VELOCITY VS. TIME AND ACCELERATION VS. TIME GRAPHS

1. The velocity vs. time graph is the most powerful of the three motion graphs, because it gives the most information about the motion. We can read the velocity right off the graph: the area under the graph is the change in position, and the slope of the graph is equal to the acceleration.

2. The area under the acceleration vs. time graph will have units of $\dfrac{m}{s^2} \times s = \dfrac{m}{s}$ meaning that the area under the graph of acceleration vs. time is equal to the change in velocity of the object during that time interval.

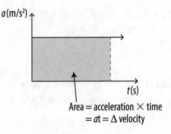

3. All three of the kinematic equations that were given earlier in the chapter can be derived from looking at areas under the curve and slopes of the position, velocity, and acceleration graphs.

4. Let's look at the table of graphs one more time and add in the acceleration graphs so that we can see how all three go together.

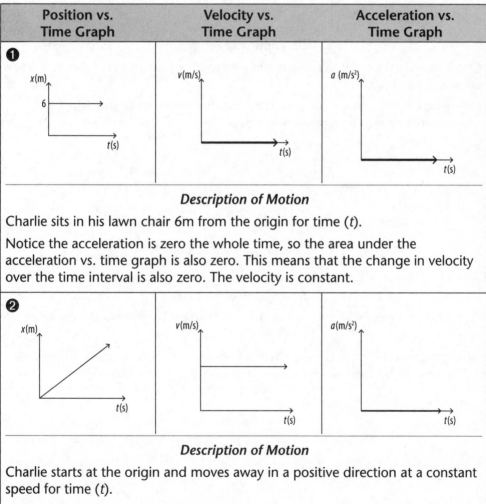

Position vs. Time Graph	Velocity vs. Time Graph	Acceleration vs. Time Graph

❶

Description of Motion

Charlie sits in his lawn chair 6m from the origin for time (t).

Notice the acceleration is zero the whole time, so the area under the acceleration vs. time graph is also zero. This means that the change in velocity over the time interval is also zero. The velocity is constant.

❷

Description of Motion

Charlie starts at the origin and moves away in a positive direction at a constant speed for time (t).

Here again the acceleration graph is zero, so the velocity is again constant.

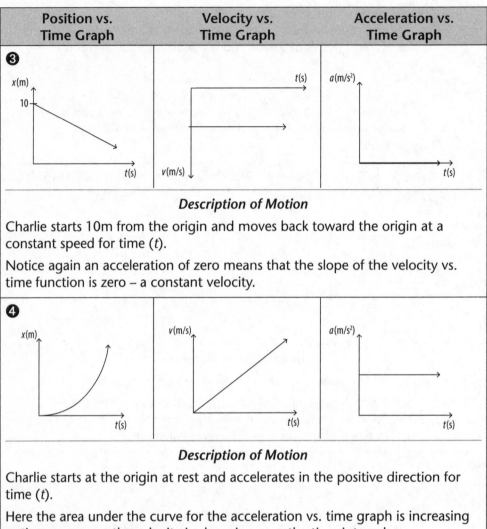

Position vs. Time Graph	Velocity vs. Time Graph	Acceleration vs. Time Graph

❸

Description of Motion

Charlie starts 10m from the origin and moves back toward the origin at a constant speed for time (t).

Notice again an acceleration of zero means that the slope of the velocity vs. time function is zero – a constant velocity.

❹

Description of Motion

Charlie starts at the origin at rest and accelerates in the positive direction for time (t).

Here the area under the curve for the acceleration vs. time graph is increasing as time passes, so the velocity is changing over the time interval.

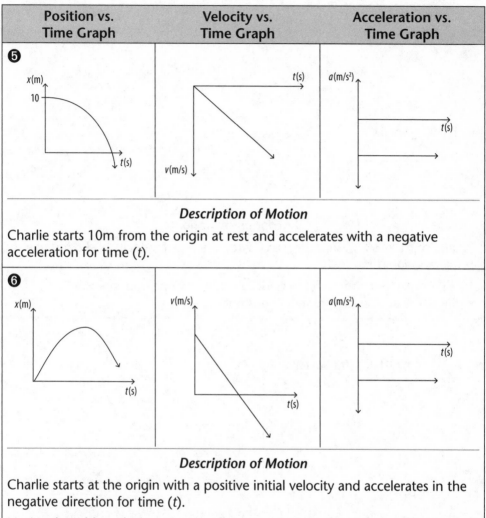

Position vs. Time Graph	Velocity vs. Time Graph	Acceleration vs. Time Graph

❺

Description of Motion

Charlie starts 10m from the origin at rest and accelerates with a negative acceleration for time (*t*).

❻

Description of Motion

Charlie starts at the origin with a positive initial velocity and accelerates in the negative direction for time (*t*).

Notice that although the velocity values change from positive to negative, the acceleration (the slope) remains constant.

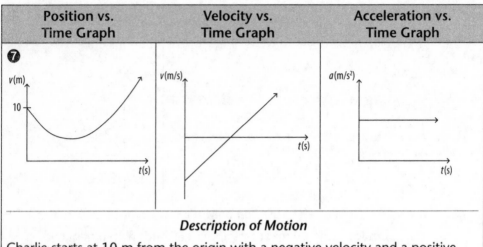

Position vs. Time Graph	Velocity vs. Time Graph	Acceleration vs. Time Graph

❼

Description of Motion

Charlie starts at 10 m from the origin with a negative velocity and a positive acceleration for time (*t*).

Again, notice that although the velocity values change from negative to positive, the acceleration remains constant.

F. GRAPHS SUMMARY

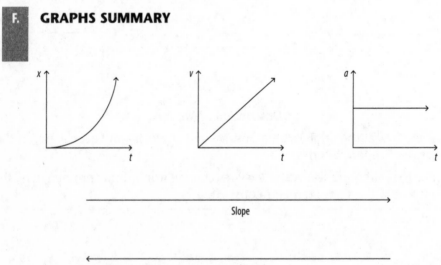

1. If you have a graph of position and need to create either velocity or acceleration, look at the slope of the graph.

2. By contrast, if you need to go in reverse order and create velocity vs. time or position vs. time graphs from an acceleration vs. time graph, look at the area under the curve.

G. USING GRAPHS TO SOLVE PROBLEMS

1. One quick example to show how you can use the velocity vs. time graph to solve a one-dimensional motion question:

  a. A rocket initially at rest is fired vertically upward from the ground and is given an initial acceleration of 2m/s^2 from the engines. The engines fire for a total of 10 seconds, after which the rocket coasts for a while before falling back down. Find the maximum height reached by the rocket.

  i. This question is relatively straightforward, except that there are two different accelerations. If you choose to solve this with equations, you have to do it in two steps.

  ii. The first step to solve this question with a graph is to sketch a velocity vs. time graph. We know that the initial acceleration is 2m/s/s (which will be the initial slope of the velocity line) until 10 seconds. After 10 seconds, we know that the rocket will be in free fall, under the influence of gravity, and have an acceleration (and hence a slope) of –10m/s/s.

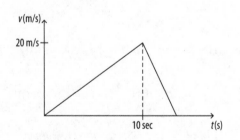

  iii. We can now label a few more things on the graph. Since we know that the initial acceleration is 2 m/s/s (or 2 m/s change in velocity for every second that passes), we know that the velocity of the rocket at 10 seconds will be 20 m/s. By the same logic, if the acceleration of the

free-fall rocket is –10 m/s/s, it should take 2 seconds for
the rocket to come to rest (which is where the rocket
will be at its highest point).

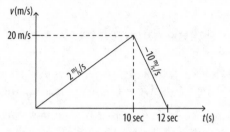

iv. To find the change in position of the rocket, look at the
area under the velocity vs. time graph. The area can be
found easily by breaking the graph into two triangles.

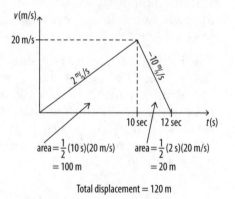

$$area = \frac{1}{2}(10\text{ s})(20\text{ m/s})$$
$$= 100\text{ m}$$

$$area = \frac{1}{2}(2\text{ s})(20\text{ m/s})$$
$$= 20\text{ m}$$

Total displacement = 120 m

v. The maximum height reached by the rocket is 120 m.

*Remember that a main point of the AP Physics 1 exam is
for you to show you understand multiple ways to represent
motion. The more ways you are able to show your knowledge
of physics, the better off you will be.*

H. REAL WORLD GRAPHS

1. On the AP Physics 1 exam, you will most likely be faced with at least one graph that looks like it comes from the real world instead of "Physics Land," where all graphs are perfect.

2. When in a laboratory, data used to produce graphs won't usually give you perfect lines. The graphs made by motion detectors contain "blips" created by the motion detector losing track of the object, or picking up your lab partner's by accident. For example, if Charlie travels away from the origin at a constant positive velocity, we might expect the position vs. time graph to look like:

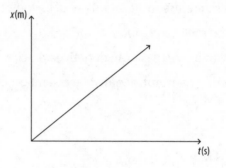

In real life and on the AP Physics 1 exam, you will be expected to look at the following and understand that you can ignore the "noise."

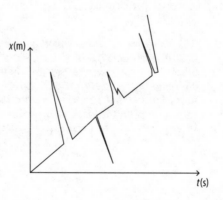

I. DESCRIBING MOTION

1. On the AP Physics 1 exam, it is likely that you will be asked to describe in words the motion of an object rather than making a direct calculation. It is important to remember that when describing the motion of an object, there are three important quantities to discuss: *position, velocity,* and *acceleration.*

2. In order to avoid missing something in your description of the motion, it is a good idea to list specifics about the position, velocity, and acceleration, rather than trying to create a paragraph. You should get to the point by indicating the direction and whether the quantity is constant, increasing, or decreasing. For example, for a rock dropped off a cliff, if asked to describe the motion of the rock, you could say:

   i.   The rock is moving downward toward the ground

   ii. With an increasingly negative velocity

   iii. And a constant negative acceleration of 9.8 m/s/s due to gravity

J. FREE FALL

1. The most common problems about one-dimensional motion seen on the AP Physics 1 test deal with objects undergoing free fall. When an object falls under the influence of gravity alone, it falls with a constant downward acceleration of 9.8 m/s^2.

Objects in free fall have only the force of gravity on them, and their acceleration is 9.8m/s/s down. There will be very few places on the AP Physics 1 exam where you will need to use a calculator. The College Board has been very clear that you **should** *use 10m/s/s for the acceleration due to gravity for any calculations on the exam. If a symbolic solution is needed, the letter* **g** *can be used for the acceleration due to gravity.*

2. For questions about *free fall*, there is a third way to think about one-dimensional motion in addition to looking at the equations and graphs. This method is especially helpful on multiple-choice questions and uses the average velocity to calculate the distance that an object has fallen.

3. To figure out how far an object has fallen, or how fast it is going after having fallen for a certain time (*t*), realize that the acceleration is constant and the acceleration is due to gravity. This acceleration of 10 m/s/s means for every second the object falls, the velocity changes by 10 m/s.

Test Tip — *You will almost always be ignoring air resistance on the AP Physics 1 exam. **Only** include it when it is exceedingly clear that the question wants you to consider it.*

4. To calculate the distance that the object has fallen, look at the average velocity. In the first second, the object isn't falling 10 m/s the whole time; instead, it starts with a velocity of zero meters per second, and at the end of the first second, it has a velocity of 10 m/s. This means that on average, the velocity of the object during the first second is 5 m/s. So if an object is traveling at 5 m/s for 1 second, it has covered a distance of 5 meters.

During the 2nd second, the average velocity is 15 m/s. So traveling at 15 m/s for one second, the object falls an additional 15 meters, and now is 20 meters from where it was dropped.

5. The chart below shows what the position, velocity, and acceleration would look like after each second for an object falling from rest off a tall cliff.

Time (Seconds)	Acceleration (m/s/s)	Velocity (m/s)	Position (m)
0	10	0	0
1	10	10	5
2	10	20	20
3	10	30	45
4	10	40	80
5	10	50	125
6	10	60	180

6. As a check, you can calculate how far the object will fall in the first 6 seconds in one step instead of stepping through it one second at a time. The average velocity for the first 6 seconds of free fall is 30 m/s. If the object is traveling 30 m/s for 6 seconds, it will cover a distance of $\dfrac{30 \text{ m}}{s} \times 6$ seconds = 180 meters.

Test Tip

The hardest part about solving a word problem is figuring out where to start, what you know, and what you need to find out. If you're totally stuck, look at units to help you assign the given quantities to the appropriate variables.

K. STEPS TO SOLVING ONE-D KINEMATICS QUESTIONS

1. The first step when solving a one-dimensional motion question is to make a list of the variables involved.

$$X_f =$$
$$X_i =$$
$$V_f =$$
$$V_i =$$
$$a =$$
$$t =$$

2. Once you've filled in the variables given from the problem, choose one of our three equations (from section A.11) to solve for the answer to the question.

3. Since you get to choose your own coordinate system, as long as you know 3 out of the 6 variables, you can solve for the rest.

Be careful—don't get too hung up on the calculations. In the AP Physics 1 curriculum, it is much more important to be able to understand and explain what is going on and why things happen, than to be able to calculate the correct numerical answer. Most of the time you'll be asked to compare or estimate values, not actually calculate an answer.

L. TWO-DIMENSIONAL MOTION: PROJECTILE MOTION

1. A *projectile* is an object launched into the air with at least some horizontal velocity, and since it is also under the influence of gravity, it will have vertical motion as well.

2. On the AP Physics 1 exam, for the most part, you will be ignoring air resistance, so the only acceleration will be 9.8m/s² in the negative vertical direction.

3. The path that a projectile takes is called a *trajectory* and is in the shape of a parabola (or a piece of a parabola, depending on the question).

4. Since there is no horizontal acceleration, the horizontal velocity remains constant throughout the projectile motion.

5. The most common two-dimensional motion on the AP Physics exam is that of objects launched horizontally.

    i. Approach this type of problem almost exactly the same as a one-dimensional question. Follow all of the same steps, but look at the horizontal motion and the vertical motion separately, as they are independent of each other.

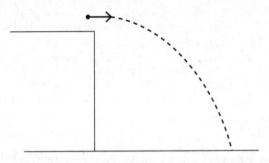

    ii. Since air resistance is being ignored, there is nothing to speed up or slow down a projectile horizontally, therefore $a_x = 0$ m/s/s. Since the acceleration in the x direction is zero, the horizontal velocity component is the same at the beginning and the end of the motion.

    iii. The vertical velocity changes in the same way as the velocity would change for a ball dropped from the top of a cliff.

    iv. As hinted at by the name, *horizontal projectile motion,* the object is launched horizontally, meaning that the initial vertical velocity is zero. This means the first equation of motion can be simplified from $y_f = y_i + v_i t + \frac{1}{2}at^2$ to $\Delta y = -\frac{1}{2}gt^2$.

v. Remember when the object lands, it will be moving both horizontally and vertically. Each of these parts is called a *component*, so an angled vector can be broken down into horizontal and vertical components using triangles and trigonometry. The final horizontal component of velocity $\left(v_{x_f}\right)$ will be the same as the initial horizontal component of the velocity $\left(v_{x_i}\right)$, and because the initial vertical velocity is zero, the final vertical component of the velocity will be equal to $v_{x_f} = -gt$.

vi. Since velocity is a vector, the total velocity will be made up of the horizontal and vertical velocities. The total, or resultant, velocity can then be calculated by using the Pythagorean theorem. The total final velocity of the object before it lands will be $\vec{v} = \sqrt{v_{y_f}{}^2 + v_{x_f}{}^2}$, and because velocity is a vector, calculation of the direction of the velocity is necessary.

vii. The direction of the velocity is usually recorded as an angle from the vertical and can be calculated with trigonometry.

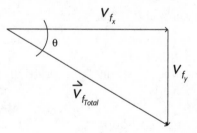

viii. The angle of the final velocity will be $\theta = \tan^{-1}\left(\dfrac{v_{y_f}}{v_{x_f}}\right)$ below the horizontal axis.

6. While horizontal projectile motion is the most common, "ground to ground" projectile motion is the simplest to visualize.

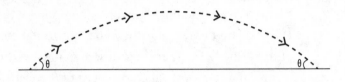

i. In a "ground to ground" case, the object has an initial velocity that has a horizontal and a vertical component. Its final vertical position is the same as its initial vertical position (no vertical displacement from launch to landing). In the case of an object launched with initial velocity (v) at angle θ, the components of the initial velocity are:

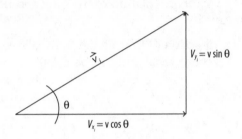

ii. If the motion is truly "ground to ground," the initial and final speeds are the same, and the angle is also the same, except now the angle is below, instead of above, the horizontal component.

7. If the projectile is being launched at an angle (either above or below the horizontal), make sure the position, velocity, and acceleration data in two dimensions (horizontal and vertical) is known before you start to fill in the charts. It is also important to note signs for direction. For example, if a snowball is launched off a tall building with an initial velocity of 20m/s at 30° above the horizontal, the velocity needs to be separated into x and y components.

i. Use trigonometry to find the initial velocity in the *x* and *y* velocity so that:

$$V_{i_x} = V_i \cos \theta \qquad\qquad V_{i_y} = V_i \sin \theta$$

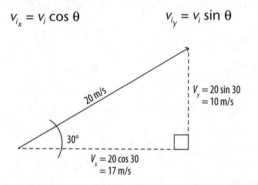

$V_y = 20 \sin 30$
$= 10$ m/s

20 m/s

30°

$V_x = 20 \cos 30$
$= 17$ m/s

Test Tip

The College Board isn't going to test your knowledge of significant figures. Remember, you are making easy calculations. Use two or three significant figures, and you should be fine.

ii. Once the information is broken into *x* and *y* components, fill in the charts and solve for anything asked.

$x_f = ?$ $y_f = 0$ m
$x_i = 0$ m $y_i = 0$ m
$V_{Fx} = 17$ m/s $V_{fy} = ?$
$V_{ix} = 17$ m/s $V_{iy} = 10$ m/s
$a_x = 0$ m/s² $a_y = -10$ m/s²
$t = ?$ $t = ?$

Test Tip

When filling out charts for vertical and horizontal motion for a projectile, the time should be the same in both directions.

iii. You should also be comfortable graphing the velocity of an object undergoing projectile motion. Remember to find the components of the projectile's velocity and graph the velocity in the horizontal and vertical directions. For example, consider a rock launched at speed (*v*) at an angle (θ) above the horizontal.

Since the horizontal component of the velocity will be constant and there is a zero horizontal acceleration, the graph of the horizontal velocity will be a constant at:

$$v_x = v \cos \theta$$

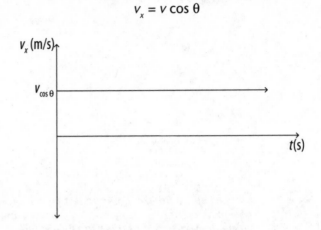

The graph of the vertical velocity should have a y-intercept equal to the initial vertical velocity of the projectile, which in the following example is equal to:

$$v_{y_i} = v \sin \theta$$

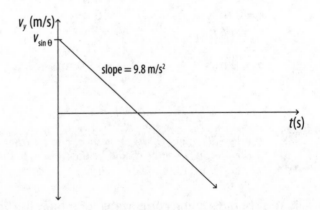

By this time, you should be able to recognize that the slope of the velocity graph should be the acceleration of the projectile, which in the vertical direction will be equal to 9.8m/s².

Newton's Laws of Motion and Force

NEWTON'S LAWS

1. *Newton's First Law of Motion* states that an object at rest stays at rest and an object in motion stays in motion, unless acted on by an outside force. This is known as the law of inertia.

2. *Newton's Second Law of Motion* states that when a *net* force acts on an object, it will accelerate at a rate which is proportional to the force and is inversely proportional to the mass of the object. The equation for Newton's Second Law of Motion is:

$$\Sigma F = ma$$

Test Tip

*Remember the equation is not just **F = ma**, but the **sum** of all the forces, or net force, that equals **ma**. If you set **any** force equal to mass × acceleration, you'll get yourself into trouble.*

3. *Newton's Third Law of Motion* states that for every action, there is an equal and opposite reaction.

   i. Newton's Third Law applies directly to forces. Another way to state Newton's Third Law is: when one object exerts a force on another object, there is a force that is equal in magnitude and opposite in direction applied by the second object on the first.

   ii. These two forces are applied by different objects and are called *action-reaction pairs*.

iii. Action-reaction pairs *always* act on two different objects and therefore do not cancel out.

iv. If a truck hits a bug, the truck and the bug both experience the exact same force. That is, the force of the truck on the bug is equal and opposite to the force of the bug on the truck. The reason that the bug is squished is because its mass is smaller than the mass of the truck. The equation for Newton's Third Law is:

$$F_{1 \text{ on } 2} = -F_{2 \text{ on } 1}$$

For the case of the truck and the bug:

$$F_{\text{Truck on Bug}} = -F_{\text{Bug on Truck}}$$

Since the force is equal in magnitude, the truck (large mass) experiences a small acceleration, while the bug (small mass) experiences a large acceleration.

$$M_{\text{Truck}} \times a = F = m_{\text{Bug}} \times a$$

B. FORCES

1. Force is measured in Newtons. 1 Newton is equal to 1 kgm/s². 1 Newton of force is needed to accelerate a 1 kg mass at 1 m/s².

2. Forces are vectors, so the rules of vector addition have to be followed. When discussing forces, directions and magnitudes matter.

Test Tip

Be careful when discussing the force of gravity in free-response questions. When discussing the force of gravity, don't just say "gravity." Remember that there is a difference between the "force of gravity" and the "acceleration due to gravity." Be sure to use the correct terminology for the situation.

i. *Push/Pull* involves direct contact. A push/pull can be caused by almost anything and is often called *applied force* (F_{Applied}).

ii. *Force of gravity* is also called the *weight* of the object and is equal to the mass of the object times the acceleration due to gravity. This force can be notated as F_g or mg.

iii. The *normal force* is a perpendicular contact force that occurs on a molecular level and acts to prevent one object from invading another object's space. The normal force will appear on any object that is in contact with a surface and will be directed perpendicularly outward from the surface. A box sitting on a table will have a force upward from the table, and since that force is perpendicular to the surface, it is called a normal force.

iv. *Tension force* is used when there are ropes or chains. Again, this force occurs on a molecular level and acts to keep that particular material in which the tension exists from pulling apart. Note if there is a pulley, the tension in all parts of the rope is the same (unless the rope or the pulley has mass; see Rotation in Chapter 9). It will, however, pull in different directions on separate objects that are attached to a given rope or chain.

v. *Spring force* occurs when there is a spring in a problem. The force that the spring applies to an object depends on how "stiff" the spring is (represented by k, the spring constant) and how far the spring is being compressed or stretched. Think about stretching a spring. If a spring is stretched by pulling on it to the right, it would apply a force to the left. The equation for the spring force for a Hooke's Law spring is:

$$F_s = -kx$$

The negative sign is a reminder the force from the spring will always be opposite to the direction of the stretch, x. Since force is measured in Newtons and meters are used to measure distance, the spring constant, k, is measured in Newtons/meter.

vi. *Friction.* Unlike the other forces, the force of friction depends on whether the object is moving or is at rest.

- *Static frictional* force is between two surfaces when the objects are at rest relative to one another. The force has a maximum threshold value, and the actual static friction force ranges from zero to this maximum threshold value. The equation for static frictional force is: $f_{static} \leq \mu_s N$, because it could equal the coefficient of friction times the normal force, but it could also be less than that value.

- μ is the coefficient of friction and depends on the two surfaces that are rubbing together. Because each pair of surfaces has a different μ value, you will not be expected to memorize any coefficients; rather you will be asked about what might happen if these values were changed. Because $\mu = \dfrac{F_f}{N}$, the coefficient of friction has no units as $\dfrac{Newtons}{Newtons}$ cancels out.

Test Tip

In most cases on the AP Physics exam, static frictional force will be calculated using the maximum value. The question will make it clear that the object is almost about to slide. If this is not the case, think carefully about what value to assign to the static frictional force. For example, if the maximum static frictional force is 10 Newtons and you are pulling on a block with a force of 5 Newtons, the frictional force must also be 5 Newtons for the block to remain at rest.

- *Kinetic friction* is the sliding frictional force between two objects that are moving relative to one another. This force is always a fixed value, which depends on the coefficient of kinetic friction and the normal force. The equation for kinetic frictional force is: $f_{Kinetic} = \mu_k N$.

C. FREE BODY DIAGRAMS

1. A free body diagram (FBD or force diagram) is used to help solve complex force problems. Unlike kinematics questions where there are 3 equations which can be applied to every situation, when dealing with force questions, you must create the equation that goes with the situation, and the FBD will help

do that. Remember forces are vectors, which have magnitude and direction. After drawing the forces, don't forget to consider components.

One of the most common errors on FBDs is drawing forces that act on objects other than the one being discussed. FBDs contain only forces ON one object, and not forces BY that object on a second object.

2. An FBD is drawn showing all the forces acting ON an object. Each object is shown as a dot, which represents the object's center of mass.

3. Each force is shown on the diagram by an arrow whose length represents the magnitude of the force and the direction of the arrow shows the direction of the force.

4. While many forces will be in the *x* or *y* direction, some may not be. These forces need to be broken into components to determine the total *x* and *y* contribution that force has to the net *x* and net *y* forces. Be careful not to draw these components on your FBD because an FBD includes the individual forces without their components. Instead, draw a separate diagram showing the components of the forces. Once all the forces in the *x* and the *y* direction are drawn, the forces must be added in the *x* and *y* directions. The sum of the forces in each direction is equal to the mass of the object times the acceleration of that object in the *x* or the *y* direction.

*Remember it is not just F = ma, but $\sum F = ma$. You must add up the forces in each direction before you set the sum equal to **ma**.*

D. COMMON FORCE SITUATIONS

1. Although it is more important to be able to draw an FBD from the situation than to memorize a diagram and an equation, it is helpful to see several common situations for which an FBD is necessary to answer a question. In the table below, you'll see several common situations with their corresponding FBDs and equations.

Situation	Illustration	FBD	Equations
Object sitting on the table		↑N • ↓mg	$\Sigma F = ma$ $N - mg = ma$ $N - mg = 0$ $N = mg$

Test Tip

Be careful here – the normal force is not the reaction force to gravity. When we discussed Newton's Third Law earlier, we said that every action has an equal and opposite reaction. Action–reaction pairs can be written: "Object A applies a force on Object B and Object B applies an equal and opposite force to Object A." There are only two objects in an action–reaction pair. For example, consider the case of a box sitting on a table. We know that the force of gravity acts on the box, so the Earth pulls down on the box. This means that the other force in the action–reaction pair is the force of the box pulling up on the Earth.

Situation	Illustration	FBD	Equations
Object sitting on the table with a second object on top		↑N • F_{b_2}↓ ↓mg	$\Sigma F = ma$ $N - F_{Block2} - mg = ma$ $N - F_{Block2} - mg = 0$ $N = mg + F_{Block2}$

Test Tip

Notice that the normal force is NOT always equal to the weight of the object. $N = mg$ is only true in select circumstances. The AP Physics 1 exam is about understanding and being able to explain, NOT about memorization.

Situation	Illustration	FBD	Equations
Object sitting on a frictionless floor with a force on it to the right		↑ N ●——→ F ↓ mg	$\Sigma F_y = ma_y$ $N - mg = ma_y$ $N - mg = 0$ $N = mg$ $\Sigma F_x = ma_x$ $F = ma_x$
Object sliding left while slowing down		↑ N ●——→ f ↓ mg	$\Sigma F_y = ma_y$ $N - mg = ma_y$ $N - mg = 0$ $N = mg$ $\Sigma F_x = ma_x$ $F = ma_x$

Test Tip

Notice that although the velocity of the object is to the left, velocity is not a force so is not shown on the FBD. The only things that should appear on an FBD are forces, and there is no such thing as a force of motion. Although it can be confusing, note that an object can be moving in the opposite direction than the net force is acting. In this case, the object is moving to the left while the friction opposes the motion and acts to the right. This causes the object to slow down because the net force (and acceleration) is opposite the direction of the velocity.

Situation	Illustration	FBD	Equations
Object sliding to the right with a constant velocity			$\Sigma F_y = ma_y$ $N - mg = ma_y$ $N - mg = 0$ $N = mg$ $\Sigma F_x = ma_x$ $F - f = m(0)$ $F = f$
Object sliding to the left while speeding up			$\Sigma F_y = ma_y$ $N - mg = ma_y$ $N - mg = 0$ $N = mg$ $\Sigma F_x = ma_x$ $F - f = ma_x$

Test Tip

Constant velocity means the velocity is not changing. If the velocity is not changing (meaning it could be either a constant positive velocity, a constant negative velocity, or zero), the acceleration is zero. If the acceleration is zero, the sum of all the forces is also zero. $\Sigma F = 0$.

Situation	Illustration	FBD	Equations
Object sliding to the right with a non-horizontal force			$\Sigma F_y = ma_y$ $N + F \sin\theta - mg = ma_y$ $N = mg - F \sin\theta$ $\Sigma F_x = ma_x$ $F \cos\theta = ma_x$

Situation	Illustration	FBD	Equations
Object moving vertically at a constant speed			$\Sigma F_y = ma_y$ $F - mg = ma_y$ (constant velocity so $a = 0$) $F - mg = 0$ $F = mg$
Object moving vertically upward while slowing down			$F - mg = -ma_y$ Or $mg - F = ma_y$

There is no force of motion showing that the velocity of the object is upward. However, since the velocity is upward, and the object is supposed to be slowing down, the acceleration has to be down. (An object slows down if its velocity and acceleration vectors point in opposite directions.) This also means the sum of the forces must point down, so the force of gravity must be bigger than the upward force.

If asked to draw an FBD, do not draw in the components for vectors. If you do so, you'll lose points on the AP Physics exam because the AP exam readers only want to see the forces acting on the object, not the components of those forces. Draw all the vectors at whatever angle you are given, and then on a different diagram, redraw the FBD with the components. This may seem silly, but it can mean the difference between 3 points and a zero.

E. NEXT STEPS

1. When dealing with forces, equations must be created. A force diagram is a tool to help create equations and to think about and visualize the forces acting on an object (or a set of objects).

2. After drawing a force diagram, analyze the direction of acceleration of the object. If the object is accelerating up and down or left and right, break up any vectors that are not horizontal or vertical into their components.

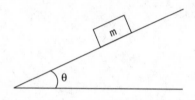

3. If the motion is not horizontal or vertical, rotate the axis so that one of the axes is pointing in the direction of the acceleration. Remember to define your coordinate system. Be consistent with the assigned directions throughout the analysis. This may seem awkward at first, but it makes the analysis much easier because rather than having to break up displacement, velocity, acceleration, and various forces into the conventional *x* and *y* directional components, you may have to break up one or two forces in the coordinate system you've chosen.

4. To rotate your coordinate system, set the x-axis so that it points in the direction of the acceleration, and the y-axis so that it is perpendicular to the direction of the acceleration.

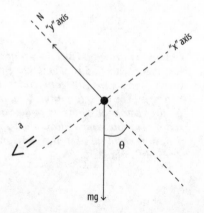

5. When finding the components of the weight, the angle between the normal force and the weight is the same as the angle of the ramp above the horizontal axis.

6. Once your axis is drawn, sum the forces in the y and the x direction and solve.

F. EXAMPLES OF FBDs

1. For an object moving to the left and slowing down, the force diagram looks like this:

(Notice that although the object is moving to the left, there is no force needed to make it move. Initially, there must have been a force, but now that the object is moving left, there no longer needs to be a force to keep it moving. There is, however, a frictional force acting in the opposite direction of the velocity, causing it to slow down.)

Now sum the forces in the y and the x directions. Let's start with the y direction:

$$\Sigma F_y = ma_y$$

The forces in the y direction are normal up, and the force of gravity is down, so our equation becomes:

$$N - mg = ma_y$$

Knowing it is not accelerating vertically, and is just sliding along the floor, one can say that $a_y = 0$ m/s², and again, our equation becomes:

$$N - mg = m(0)$$
$$N - mg = 0$$
$$N = mg$$

Now sum the horizontal forces.

$$\Sigma F_x = ma_x$$
$$-f = ma_x$$

Once you have reached this step, any question should be simple to answer using the equations you have created. Again, every situation will be slightly different. However, by drawing an FBD, analyzing all the forces in the x and y directions, and remembering that it is the *sum* of all the forces in those respective directions, you will be successful when analyzing force problems.

2. The *Atwood machine* is a common situation likely to be on the AP Physics 1 exam. A basic Atwood machine is two blocks connected by a string over a pulley. The two blocks (of different masses) accelerate together, and the acceleration and the tension in the string can be calculated if the masses of the two blocks are known.

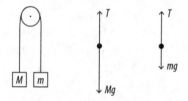

Since there are two masses, two FBDs are necessary for the situation. Once the two FBDs are drawn, sum the forces for each block and solve. Solve for what? Anything. No matter what is asked, follow the same steps to solve the FBD.

Left Block	Right Block
$\sum F = Ma$	$\sum F = ma$
$T - Mg = -Ma$	$T - mg = ma$
Notice the addition of a negative sign for the acceleration of the larger block. Forgetting this step means calculating an acceleration of 10 m/s/s, which cannot be correct since it is tethered to another block. It is better to think carefully and slowly and to remember this block is heavier and so will have a downward acceleration.	The tension in the string and the acceleration of the blocks is not known, but we do know the blocks must be accelerating together. How do we know this? Well, they are attached, and how strange the world would be if the two blocks didn't have the same acceleration! Also, we know that since they are connected with one cord, the tensions in the two equations must also be equal.

Now there are two equations and two unknowns. The equations can be solved algebraically for the tension in the cord or the acceleration of the blocks.

Atwood machines can get more complicated. They could put one of the masses on a surface of a table,

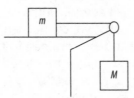

Or they could put one on an angled table.

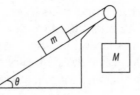

However, once able to draw FBDs, find components, and use Newton's Second Law ($\sum F = ma$), nothing can trip you up.

G. **MASS VS. WEIGHT**

1. Mass is a measure of how much matter an object contains. Mass is measured in kilograms (kg). There are two types of mass: inertial and gravitational.

   i. Inertial mass is the property of mass that resists changes in an object's motion. An object with a large inertial mass will have a large inertia—meaning the object will resist changes in its motion more than an object with a smaller inertial mass. Inertial mass can be calculated by applying a known force to an unknown mass and measuring the acceleration. Newton's Second Law can then be applied, and the inertial mass will be $m = F/a$.

   Inertial mass is usually measured by making observations of an object that is oscillating. (This idea will be explored further in Chapter 10: Simple Harmonic Motion.)

   ii. Gravitational mass is the property of mass that responds to gravitational forces. Gravitation is a field force, and matter is the property affected by this particular field. This same idea is behind other field forces, such as the electrostatic force and the magnetic force. The electrostatic field affects anything that has a charge, while the magnetic field affects anything that is magnetic. Gravitation works in the same way but is more difficult to visualize because while this field affects anything with mass, it is usually on such a large scale that we cannot see the masses' behavior within that field and are confined to the surface of the Earth. The gravitational mass is measured by comparing the force of gravity of an unknown mass to the force of gravity of a known mass. Gravitational mass is usually measured with a balance or a scale.

   iii. When discussing the mass of an object, do not distinguish whether referring to its inertial mass or its gravitational mass. The quantity of matter (its mass) does not depend upon the method by which it is measured; however, a conceptual understanding of the difference between the gravitational and inertial masses is needed.

2. Weight is the measure of the force of gravity between two objects. The weight of an object can change depending on where the object is. For example, an elephant will have a different weight at sea level than it will at the top of Mt. Kilimanjaro because the gravitational field (g) is slightly different due to the varying radius as measured from the center of the Earth. (A small difference, but still a difference!) The same elephant will have a very different weight on the Moon or on Mars than it would on the surface of the Earth because these objects have a very different mass and radius than the Earth and, therefore, different gravitational fields.

3. Weight is measured in Newtons and can be labeled F_g, mg, W, or Weight.

Test Tip

Be sure to know and be able to explain the difference between mass and weight!

Circular Motion and Gravitation

 A. **CIRCULAR MOTION**

1. Rotational Motion

   a. Objects experiencing uniform circular motion travel in a circle at a constant *speed*.

   b. If an object makes a complete circle, the distance traveled is the circumference of that circle or $2\pi r$. The speed of the object can be calculated by $v = \dfrac{2\pi r}{T}$, where T is the period or the time it takes for one complete revolution or cycle measured in seconds.

   c. The direction of the instantaneous velocity is always tangent to the circle (and is sometimes called tangential velocity). The direction of the velocity is constantly changing; therefore, objects in uniform circular motion do not have a constant velocity (even if the speed is constant). The magnitude of this velocity is referred to as tangential speed and is usually constant. If asked about the motion of an object at a specific point, you would need to describe the tangential velocity and consider the magnitude and direction of the velocity at that point.

   d. Remember acceleration is a change in velocity, and because the tangential velocity is not constant (the direction is always changing), objects in uniform circular motion are constantly accelerating. The acceleration of an object traveling in a circle is called centripetal acceleration.

*Centripetal force is **not** a new force. It is the name given to the **net** force in a situation where an object is traveling in a circle. Sometimes the centripetal force will be tension, friction, the force of gravity, the normal force, or a combination of these or other forces.*

e. Centripetal acceleration is a vector and always points toward the center of the circle, or radially inward. The magnitude of the centripetal acceleration is $a_c = \dfrac{v^2}{r}$, where v is the speed of the object, and r is the radius of the circle.

f. If an object is undergoing uniform circular motion, the centripetal acceleration is caused by the net force, which can also be labeled the centripetal force, and *both* point toward the center of the circle. Because this force (and acceleration) are perpendicular to the velocity and have no components in alignment with the velocity, they act only to change the direction of the motion (causing it to constantly turn or go in a circle) rather than causing it to speed up.

*Try not to get tripped up by the direction of the centripetal acceleration and net force. Your intuition might lead you to believe that the force and the acceleration point outward, away from the center of the circle, which would be called a "centrifugal" force and acceleration. Be careful! There is **never** a centrifugal force or acceleration. This confusion comes from our experience with inertia. You may have ridden in a car rounding a sharp corner and felt as if you were being pushed toward the outside of the corner. In this example, there is no force pushing you outward. You are just feeling your inertia. Remember all objects with mass have inertia and tend to resist changes in their motion. So while friction between your bottom and the seat pulls you in toward the center of the circle (the centripetal force), you feel your inertia, which makes you feel like you want to continue in a straight line.*

g. Solving problems for objects in uniform circular motion involves the same strategies employed to solve other questions with forces. Draw a free-body diagram; determine the net force in both the *x* and *y* directions, and set the net force equal to *ma*. Allow the acceleration to equal the centripetal acceleration when you are summing the forces contributing to the circular motion of the object. (Keep in mind these could be *x* or *y* forces depending on the situation.)

Situation & Illustration	FBD	Equations
Ball being swung on a horizontal string of length ℓ on a frictionless table.		The ball is not accelerating in the vertical direction, so the net force is equal to the tension in the string. Tension is the centripetal force. $$\sum F = T = ma_c$$ $$T = \frac{mv^2}{\ell}$$
Ball being swung on a vertical string.		At the top of the vertical loop, the tension and the gravitational forces are acting downward (toward the center of the circle). $$\sum F = T + mg = ma_c$$
Ball being swung in a conical pendulum.		The ball is traveling in a horizontal circle, so there must be a net force toward the center of the circle. $$\sum F_y = T\sin\theta - mg = 0$$ $$\sum F_x = T\cos\theta = ma_c$$
Car on a banked curve of angle θ.		Even though the car is going in a circle on a banked ramp, the car is still making a horizontal circle, so the net force must still be horizontal. $$\sum F_y = N\sin\theta - mg = 0$$ $$\sum F_x = N\cos\theta = ma_c$$

Situation & Illustration	FBD	Equations
Car going over a small hill, maximum speed.		Even though the car is not making a complete circle, the bump at the top of the hill is a piece of a circle. At the top of the hill (the most likely place for the car to lose contact with the ground), the forces are $$\sum F = N - mg = -ma_c$$ If the car goes too fast, the car will lose contact with the road, making the normal force equal to zero. So the *maximum* speed the car can go without losing contact with the road is: $$\sum F = N - mg = -m\frac{v^2}{r}$$ $$0 - mg = -m\frac{v^2}{r}$$ $$v_{max} = \sqrt{gr}$$
Car going through a vertical loop-de-loop, minimum speed.		The most likely place for the car to fall off the track is again at the top. $$\sum F_y = -N - mg = -ma_c$$ This time, if the car goes too slow, the car will come off the track, making the normal force equal to zero. So the *minimum* speed the car can go without losing contact with the track is: $$\sum F = -N - mg = -m\frac{v^2}{r}$$ $$\sum F = 0 - mg = -m\frac{v^2}{r}$$ $$v_{min} = \sqrt{gr}$$

B. GRAVITATIONAL FORCE

1. The gravitational force between two objects with mass can be calculated with the equation $F_{gravity} = \dfrac{GM_1 M_2}{R^2}$, also known as Newton's Law of Universal Gravitation.

Newton's Third Law still applies here! The Earth and the moon exert a gravitational force on each other. The force exerted by the Earth on the Moon is equal to and opposite in direction to the force exerted by the Moon on the Earth.

2. R is measured as the distance from center of mass to center of mass of each object, and G is the universal gravitational constant and is equal to $F_{gravity}$.

*The "R" in the gravitational field equation is **not** the radius of the planet. It is the distance from the center of the planet to the place in space that you are investigating.*

3. The force of gravity between a planet and an object is also called the object's weight, so $\dfrac{GM_1 M_2}{R^2} = M_2 g$.

On the AP Physics exam, it is unlikely you will have to calculate the actual acceleration due to gravity on a different planet. However, you might be asked to estimate the acceleration due to gravity on a different planet, in terms of g the acceleration due to gravity on the Earth. For example:

Planet X is half as massive as the Earth and has twice the radius of the Earth. What is the acceleration due to gravity on planet X in terms of "g," the acceleration due to gravity on the Earth?

$$F_g = M_2 g = \frac{GM_1 M_2}{R^2}$$

$$g = \frac{GM_1}{R^2}$$

$$g_{Planet\ x} = \frac{G\left(\frac{1}{2}M_E\right)}{(2R_E)^2} = \frac{1}{4} \times \frac{GM_E}{R_E^2}$$

$$g_{Planet\ x} = \frac{1}{8}g_{Earth}$$

C. GRAVITATIONAL FIELD

1. Gravity is a non-contact force that acts between two objects even if they are some distance apart. This is tricky. How can the Earth exert a force on something that it is not touching? The best explanation to this question involves the idea of the gravitational field.

2. The gravitational field is a vector field that describes how a massive object affects the space around it. Other masses in that field would "feel" the alteration of the space, but there does not have to be a second object for there to be a gravitational field.

3. The acceleration due to gravity, *g*, is equal to the magnitude of the gravitational field produced by an object and is equal to:

$$\overline{g} = \frac{GM_1}{R^2}$$

4. This is measured in Newtons/kg or m/s^2.

D. GRAVITATIONAL FORCE AND ELECTROSTATIC FORCE

1. There are two fundamental non-contact forces that you will encounter on the AP Physics 1 exam: the gravitational force and the electrostatic force.

2. The gravitational force and the electrostatic force are action-at-a-distance forces (which is another way of saying non-contact forces), and their equations are very similar:

$$F_{gravity} = \frac{GM_1M_2}{R^2} \text{ vs. } F_{electrostatic} = \frac{KQ_1Q_2}{R^2} .$$

3. Notice that both of these forces are directly related to the objects involved in the interaction. (In the case of gravitation, this is mass [*m*] and in the case of the electrostatic force charge [*Q*]). Both equations involve a constant (*G* and *k*), and finally, both forces follow an inverse square relationship with respect to the distance between the two objects involved in the interaction. This means that this force decreases as the distance increases. And because it is an inverse square relationship, it decreases at a very fast rate. If the two objects are a distance (*d*) apart, and then moved to a distance of 2*d* apart, the force decreases by a factor of $\frac{1}{4}$ (so halving the distance changes it by a factor of $\frac{1}{4}$, not $\frac{1}{2}$). If the distance were changed to 3*d*, the force would change by a factor of $\frac{1}{9}$; change the distance to 4*d*, and the force would be $\frac{1}{16}$, and so on.

4. The electrostatic force (see Chapter 12) is a stronger force than the gravitational force, yet it is the gravitational force that seems to dominate our everyday experience. There are two reasons for this observation.

   a. The gravitational force seems much stronger because the Earth is massive.

   b. All mass creates a gravitational field and exerts gravitational forces on other masses in the neighborhood. Mass is only ever positive, so all gravitational forces are attractive. Charges can be positive or negative and can create an attractive or a repulsive electrostatic force. Because the Earth is generally electrically neutral, the net effect of all the electrostatic forces is close to zero.

Work, Energy, and Power

WORK

1. When a force is applied to an object and that object moves parallel to the direction of the force, we say that work is done on that object.

2. The equation for work is $W = Fd\cos\theta$. F is the force, d is the displacement of the object, and θ is the angle between the force vector and the displacement vector. The $\cos\theta$ term means that when the force and displacement are in parallel or anti-parallel with one another, the amount of work done will be at a maximum as $\cos(0) = 1$. It also means that no work will be done when the force is perpendicular to the displacement, as $\cos(90) = 0$. There can still be work done on an object when the force is partially aligned with the displacement, at an angle somewhere between 0° and 90°; however, it will not be at a maximum because $\cos\theta$ will range between 0 and 1 for these angles. Note also that the angle between the force and displacement could be anti-parallel (opposite in direction). In this case $\theta = 180°$ and because $\cos(180°) = -1$, we get a negative value for the work. Work is a scalar, so this is not giving us the direction of the work; rather, it tells us this force is decreasing the energy of the object rather than increasing the energy. Friction is a force that often does negative work on moving objects, thereby acting to decrease their energy.

Example & Diagram	Free-Body Diagram	Work Done by Force *F*
An object is pulled to the right a distance (d) across a level, frictionless surface by a horizontal force.	N $\uparrow$ $\bullet \longrightarrow F$ $\downarrow mg$	$W_F = Fd\cos(\theta) = Fd$

Test Tip

Notice the gravitational force and the normal force do not do work on the object in this first example because the displacement of the object is perpendicular to the normal and the gravitational force.

Example & Diagram	Free-Body Diagram	Work Done by Force *F*
An object is pulled to the right a distance (d) across a level, frictionless surface by a force directed at an angle θ to the horizontal.	N $\uparrow$ $\nearrow F$ θ $\downarrow mg$	$W_{Force} = Fd\cos\theta$
An object is pulled upward a distance (d) at a constant speed by a force (F).	$\uparrow F$ $\bullet$ $\downarrow mg$	$W_{Force} = Fd\cos\theta = Fd$

Example & Diagram	Free-Body Diagram	Work Done by Force *F*
An object is displaced up an incline a distance *d* at a constant speed with force (F). 	N F ↓ mg	$W_{Force} = Fd\cos\theta = Fd$
An object is swung in a horizontal circle on a frictionless table by a force (F).	↓ F	$W_{Force} = Fd\cos90 = 0$

Test Tip

*Remember that the angle θ is the angle **between** the force vector and the displacement vector. The AP Physics 1 exam will test your understanding of this by giving you an angle. Resist the temptation to plug and chug! You have to THINK!*

3. If you forget the units for work, you can figure it out by looking at the quantities used to calculate the work. Remember work is the product of force and displacement, so the units of work should be Newtons times meters. One Newton meter is also called a Joule.

$$1Nm = 1 \text{ Joule}$$

4. Even though force and displacement are vectors, the product of force and displacement is a scalar. Work can be positive or negative but it never has a direction. The positive and negative tell us whether energy is being added to or taken out of a system.

5. Finding the net work

It is possible to find the net work in two ways.

a. Method 1—Find the net force by using a free-body diagram. Once you know the net force, find the component of the net force that is parallel to the displacement and multiply that component of the force by the displacement.

b. Method 2—Find the work done by each individual force. Then sum the works done, paying special attention to the signs of each work. Remember work can be positive or negative.

6. Graphs of force vs. displacement

a. Graphs of net force vs. displacement can be used to find the work done on an object. Work is the product of net force and displacement, so the work will be the area under the net force vs. displacement graph.

b. If you are given a graph of net force vs. displacement and asked to calculate the work, find the area between the function and the axis, just like with graphs of velocity vs. time.

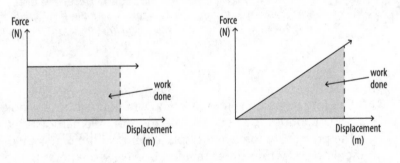

B. ENERGY

1. Energy is the ability to do work.

   a. Kinetic energy is the energy of motion.

      i. An object will have *translational kinetic energy* if the center of mass of the object is moving. Translational kinetic energy can be calculated with the equation $KE = \frac{1}{2}mv^2$.

      ii. An object that is rotating will have rotational kinetic energy. Rotational kinetic energy can be calculated with the equation: $KE_{Rot} = \frac{1}{2}I\omega^2$. (The use and application of this equation will be discussed in Chapter 9.)

      iii. Kinetic energy is measured in Joules. $1\ J = kg\frac{m^2}{s^2}$.

   b. Potential energy is the energy of position of objects in a system. When working with potential energy, remember the first step is to choose a position to be zero. When objects are above or below (or in the case of springs, potentially left of or right of) that position, they will have potential energy. It is important to remember that when you lift a book off a table, it doesn't have gravitational potential energy. Rather the gravitational potential energy is stored as a result of the positions of the system.

      i. There are two types of potential energy: gravitational and spring.

         • The equation for gravitational potential energy is $U_g = mgh$.

Test Tip

On the AP Physics 1 equation sheet, the gravitational potential energy equation is also written $U_G = -\frac{Gm_1m_2}{r}$.

*Remember **r** is not a radius measurement, but rather a distance measured from center of mass to center of mass between two objects (usually a planet and another object) for which the potential energy is being calculated.*

- If the object is at the surface of the Earth, the equation U_g= mgh will be all you need, since the value of "g," the acceleration due to gravity is relatively constant over the distances in question. If the object of interest is significantly above the surface of the Earth (for example, if you are studying the motion of the Moon with energy), use a more general form of the gravitational potential energy:

$$U_g = -\frac{Gm_1 m_2}{r}$$

- In this equation, G is the universal gravitational constant:

$$G = 6.67 \times 10^{-11} \frac{Nm^2}{kg^2}$$

- Using this equation, we can find the potential energy of the system—including two masses, where r is the distance between the two masses. (The negative sign comes from the fact that in order for us to work with potential energy we need a reference point. The convention is to choose that reference point infinitely far away and call that zero potential energy. Potential energy is the work done against the force of gravity to move an object from that zero reference point. The force is opposite in direction, which is where the negative sign comes from.)

ii. Spring potential energy is the energy stored in a spring when it is stretched or compressed.

- The force exerted by a Hooke's Law spring on an object varies with position:

$$F = -kx.$$

(Because the force is not constant, the only way to deal with a spring is with energy.) k is the spring constant and represents the stiffness of the spring, the higher the value of k, the stiffer the spring. k is measured in Newtons/meter (N/m). The negative sign in the force equation is a reminder that the direction of the force exerted by the spring is opposite to the direction of the spring's displacement.

Test Tip

The force exerted by a spring depends on the distance that the spring is stretched or compressed. This means the force exerted by a spring is not constant, so the acceleration of a mass attached to a spring is not constant. Because of this, kinematics cannot be used to solve for position or velocity of a mass on a spring – energy must be used. See a spring? Think energy!

- Every spring has an equilibrium position. This represents the naturally unstretched or uncompressed position of the spring.

- A spring can also reach equilibrium when there are forces acting on it; in this case, it will be stretched out or compressed, but the net force is zero. Equilibrium can be useful for determining the spring constant of the spring. For example, when an object of mass (m) is hanging from a spring, the object is at equilibrium when the spring force upward is equal to the gravitational force downward. To find the equilibrium position, sum the forces on the mass:

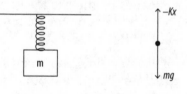

$$\sum F = kx_{equ} - mg = ma$$

- When the mass is at equilibrium, the acceleration equals zero and the equation can be rewritten:

$$kx_{equ} = mg$$

$$x_{equ} = \frac{mg}{k}$$

- The potential energy stored in the spring depends not only on how far the spring has been compressed or stretched, but also on k, the spring constant. The spring constant tells us how stiff the spring is. The spring constant is measured in Newtons/meter.

- Spring potential energy can be calculated with the equation

$$U_s = \frac{1}{2} kx^2$$

- The zero point for spring potential energy is the unstretched or uncompressed equilibrium position of the spring.

C. CONSERVATION OF ENERGY

1. The Law of Conservation of Energy states energy cannot be created or destroyed, but it can be transformed from one kind of energy to another and be transferred between objects.

2. There are two types of forces: conservative and non-conservative.

   a. When conservative forces do work on an object, the work done depends on the change in position (displacement) of the object, and not on the path the object took (distance traveled). This is called path-independence.

      i. For example, imagine boxes sliding down two frictionless ramps. In both cases, the work done by gravity is W = mgh. The only thing that matters is the height of the ramp (which is the vertical displacement of the box), not the length of the ramp.

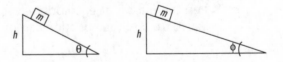

b. When non-conservative forces do work on an object, the work done depends on the path the object takes. In other words, the work done by non-conservative forces is path-dependent.

   i. For example, using the box example above, if the ramps now are frictional, the work done by friction will not be the same in the two cases, since the frictional work depends on the length of the ramp.

   ii. Most of the other named forces will be non-conservative forces.

D. WORK-ENERGY PRINCIPLE

1. The work-energy principle states that work done by the *net* force on the system is equal to the change in energy of the system. While this is true, it is *not* true that *every* force acting on an object changes the energy of the system. As mentioned, it is easier to think about the work done by the non-conservative forces only and write the work-energy theorem as

$$W_{NC} = \Delta E$$

2. Pay attention to what you are being asked because it could be the work done by a specific force or the net work done. In many cases, the net work done is zero, but this means that while one force was doing positive work on an object, another force was doing negative work on the object.

E. PROBLEM SOLVING WITH WORK AND ENERGY

1. There are only two steps to solving any question involving work and energy:

   a. The first step is to determine the forces acting on the object or system in question. (See Chapter 5 for a discussion of force diagrams.)

   b. Once the force diagram is drawn and you know what force or forces will be doing work on an object, decide if the forces that are doing work are conservative or non-conservative forces.

   c. If they are conservative forces only, use the conservation of energy to solve the problem. If there are non-conservative forces doing work, use the work-energy theorem.

Test Tip

> *Besides being able to use the work-energy principle to solve problems, a good conceptual understanding of work and energy is needed. The questions on the AP Physics 1 exam will likely test your ability to explain when and how you would use the work-energy principle to solve a problem.*

2. Examples

   a. Example 1. A box sliding along with an initial velocity (v_o) on a rough table (μ_k) comes to rest a distance (d) from where it was pushed. Derive an expression for the initial velocity in terms of v_o, μ_k, and d.

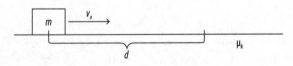

   i. To solve, decide if there are conservative or non-conservative forces at work.

ii. There are three forces acting on the box: normal, gravitational, and frictional. The normal and the gravitational forces are perpendicular to the displacement of the box, and don't do any work. The frictional force is a non-conservative force, so the work-energy principle will be used to solve this question.

iii. Work = ΔE

$$F d \cos\theta = E_f - E_o$$

$$F_k d \cos(180) = 0 - KE_o$$

$$-fd = -\frac{1}{2} m v_o{}^2$$

$$\mu_k mgd = \frac{1}{2} m v_o^2$$

$$v_o = \sqrt{2\mu_k gd}$$

In this case, the friction force is doing the work. Remember $f_k = \mu_k N$.

b. Example 2. A student pushes an object attached to a spring a distance *x*. How much work does the student do to compress the spring?

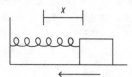

i. Be careful. You might be tempted to use the equation $W = Fd\cos\theta$. But that equation is only good for constant forces. Because the force that the student is exerting on the object changes as he pushes the object further in, you must use energy to analyze this problem.

ii. We know Work = ΔE, so if we can find the change in the energy of the spring-box system, it must be equal to the work done on the box.

iii. The initial energy of the box (assuming that it was at equilibrium) is zero. The final energy of the box is all spring potential ($U_s = \frac{1}{2} kx^2$). (The box could also have gravitational potential energy, depending on where you chose zero height to be, but because it is sliding across the table, and not changing height, it will have the same gravitational potential energy at the end as it did in the beginning.)

$$W = \Delta E$$
$$W = E_f - E_o$$
$$W = \frac{1}{2} kx^2 - 0$$

iv. The work done by the student in compressing the spring is equal to the energy stored in the spring.

$$W = \frac{1}{2} kx^2$$

F. ENERGY DIAGRAMS

1. An energy bar chart is a tool used to show understanding of the work-energy theorem. In a work-energy bar chart, a bar is drawn for each form of energy, and the sum of all the types of energy plus the work done on the object by the external forces equals the sum of all the final energies.

$$KE_i + U_i + W_{ext} = KE_f + U_f$$

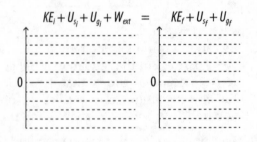

2. The exact heights of the bars are not important, but the relative heights of the bars are.

3. For example, when a box of mass (*m*) is released from rest and slides down a frictionless incline, the energy diagram looks like this:

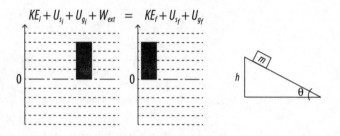

$$KE_i + U_{s_i} + U_{g_i} + W_{ext} = KE_f + U_{s_f} + U_{g_f}$$

Initially, the energy is all gravitational potential energy. Since the incline is frictionless, there are no non-conservative forces acting on the box, so the final energy should be equal to the initial energy. The final energy of the box is kinetic.

4. If the incline was not frictionless, the bar chart would have to include the work done by friction. Because the work done by friction will be negative, the bar for work will be below the zero line.

$$KE_i + U_{s_i} + U_{g_i} + W_{ext} = KE_f + U_{s_f} + U_{g_f}$$

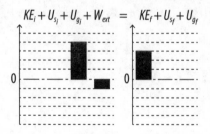

You will likely find questions on the AP Physics 1 exam asking you to sketch an energy bar chart instead of solving for a final answer. For example: a box, initially traveling at speed (v), is pushed in the direction of motion by a force (F) for time (t). After time (t), the box has a speed (2v). Sketch an energy diagram for the box.

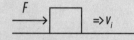

Notice you are not asked to specifically solve for the final kinetic energy of the box, nor are you asked to solve for the work done on the box. But, in order to sketch the bar chart accurately, think about what these two quantities will be equal to.

*Since you know that the speed at the end is twice the original speed, you can calculate that the final kinetic energy is **four** times larger than the original kinetic energy because kinetic energy (KE) is proportional to (v) squared. So if the initial kinetic energy is 1 unit high, the final kinetic energy is 4 units high. You also know that enough work must have been done so that the initial kinetic energy plus the work equals the final kinetic energy.*

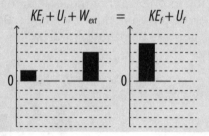

G. POWER

1. Power is the rate at which work is done or the rate of change of energy.

2. $P = \dfrac{\text{Work}}{\text{Time}} = \dfrac{\Delta\text{Energy}}{\text{Time}}$

3. Power is measured in Joules/second or Watts.

4. If a net force is causing an object to have a constant velocity, power can be calculated by using the equation:

$$P = \frac{\text{Work}}{\text{Time}} = \frac{F_{parallel}D}{\text{Time}}$$

We know that displacement/ time = velocity, so this equation can be rewritten as:

$$P = Fv$$

Test Tip

Remember P = Fv is only for situations where the object has a constant velocity. If the velocity is changing, go back to Power = Work/time.

5. Be sure you know the difference between work and power. For example, if you are asked to push two identical boxes up two different frictionless ramps as shown, the work done on the box in each case will be the same.

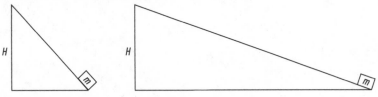

a. The work done on the box will be equal to the change in gravitational potential energy.

$$\text{Work} = mgH$$

The shorter, steeper ramp would seem more difficult to push the box up; however, you would have to push it a much shorter distance. While the longer ramp would be much easier to push the box up, you would have to push it much further.

$$\text{Work} = \text{Force} \times \text{Displacement}$$

(FORCE) × (DISPLACEMENT) = (FORCE) × (DISPLACEMENT)

b. Even though the amount of work done on the boxes would be the same, the power required would be different for each box and would depend on the amount of time required to push each block to the top of the ramp. In this situation, assuming the blocks were pushed at the same speed, it would take longer to get to the top of the longer ramp. Therefore, the power required for the longer ramp would be *less* than the power required for the steeper ramp.

Momentum

A. MOMENTUM

1. Momentum is a measure of an object's motion. The amount of momentum an object has depends on the mass and the velocity of the object. Unlike other quantities, momentum does not have a special name for its unit. The units of momentum are units of mass times velocity or kg·m/s. The equation for momentum can be written as:

$$p = mv$$

2. Momentum is represented by the symbol p. (Be careful you don't confuse this with the symbol P that stands for power.)

3. Because momentum is the product of a scalar quantity (mass) multiplied by a vector quantity (velocity), it is also a vector. Therefore, it has both magnitude and direction. Be careful about direction when analyzing the momentum of an object.

4. Momentum is useful in the analysis of moving objects because it, like energy, is often conserved. Momentum is conserved if there is no external force on the system.

Test Tip

Whenever a problem on the AP Physics 1 exam involves a collision or explosion, first try to solve the problem using conservation of momentum before using forces or energy.

B. DEFINING A SYSTEM

1. The first step in using momentum to analyze a problem is to *define your system*.

> **Test Tip**
>
> *The idea of systems will be **very** important on the AP Physics 1 exam. You will be asked to define your system and justify your choice, so be prepared to think critically about it.*

2. If we want to use conservation of momentum, remember that momentum is conserved *only if there are no external forces on the system*. Therefore, choose the system to include all the objects that might exert forces.

   Example: Two carts are traveling toward each other on a frictionless table. They will soon collide, so we should use momentum to analyze the situation. What should we choose as our system?

   The table is frictionless, so we don't have to include the table in our system, and we'll ignore air resistance, so we can keep the atmosphere out of our system. That leaves the two carts. If we include only one cart in our system, the second cart will collide with the first and exert an external force (a force from outside the system), so momentum won't be conserved.

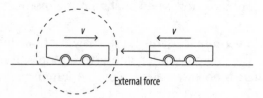

External force

   If, instead, we choose *both* carts to be our system, the only force is the force between the two carts, which is *inside* the system and, in this case, momentum will be conserved within our system. Analyzing the problem this way is simpler.

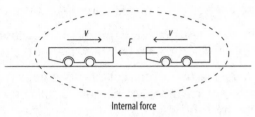

Internal force

3. Once you've defined your system, you're ready to analyze the situation with momentum.

Test Tip *In our example, we didn't include the table in our system because it was frictionless. Even if the table was rough, we could still **exclude** it from our system. The effect of the frictional force during the collision (from **just before** to **just after**) will be negligible for situations you are asked to analyze on the AP Physics 1 exam.*

C. CONSERVATION OF MOMENTUM

1. Once your system is defined and there are no external forces, use the equation for the conservation of momentum. Remember the law of conservation of momentum says that if there are no external forces on the system, the momentum is conserved, which means that the momentum before and after the collision will be the same. Write that out as an equation:

$$p_o = p_f$$

Because there were two objects, add the momentums from each object before and after the collision.

$$m_1 v_{1_i} + m_2 v_{2_i} = m_1 v_{1_f} + m_2 v_{2_f}$$

Test Tip: *When is a collision not a collision? Explosions, or situations in which two things are touching and then push each other apart, are collisions in reverse. The same rules apply as in a collision.*

2. There are two types of collisions: inelastic and elastic.

a. In an inelastic collision, momentum is conserved, and objects are separate before and after the collision.

b. In a *perfectly inelastic* collision, the objects stick together after the collision. Momentum is still conserved, and because the two objects are stuck together, they will have the same final velocity after the collision occurs, so in the momentum equation above, $v_{1_f} - v_{2_f}$. Perfectly inelastic collisions can also be called totally inelastic or completely inelastic and can also include explosions. Think of an explosion as an inelastic collision in reverse, where the objects start as one object and then are separate after the explosion occurs. It may be confusing to think that momentum is conserved in a situation like this. If an object is not moving (and therefore has zero momentum), then explodes into various pieces all of which are moving (and each piece has momentum) how can it be conserved? The key is to remember that momentum is a vector quantity. Although each piece has momentum, some will have momentum in the positive direction and some in the negative direction. When you add all the momenta together, it adds to zero, so the system will still have a net momentum of zero. Kinetic energy is not conserved in this type of collision; it usually is transferred into non-mechanical forms, such as heat and sound, and also goes into deforming the objects.

c. In *elastic collisions,* momentum *and* kinetic energy are conserved, so the conservation of momentum equation and a similar equation for the conservation of kinetic energy will be set up:

$$KE_i = KE_f$$

$$\frac{1}{2}m_1 v_{1_i}^2 + \frac{1}{2}m_2 v_{2_i}^2 = \frac{1}{2}m_1 v_{1_f}^2 + \frac{1}{2}m_2 v_{2_f}^2$$

d. It is possible that for an elastic collision, you would need to find the final velocities of each cart, which would mean that you would need to solve simultaneous equations.

Most likely you would have to determine *if* a collision was elastic. Kinetic energy is conserved only in elastic collisions; therefore, if you are asked if a collision is elastic, check the initial and final kinetic energies. If the kinetic energies before and after are the same, you have an elastic collision. If the final kinetic energy is different than the initial kinetic energy, the collision is *inelastic*, and the kinetic energy must have been converted to other kinds of energy, such as heat or sound, or work done by a non-conservative force exerted by one object on the other.

3. Two-dimensional collisions

Test Tip

While it is not likely that you will be asked to solve a numerical problem involving a two-dimensional collision on the AP Physics 1 exam, it is important you understand the concept and how to solve for what is asked.

The most likely set of questions involving a two-dimensional collision would revolve around being able to understand and explain that because momentum is conserved, certain things (such as the y-components of momentum adding to zero) will result from a collision of two objects.

a. If the collision between two objects moving in a straight line toward each other is a glancing collision, meaning that it is not head on center to center, then the two objects will not continue to move in a straight line after the collision.

b. The collision of two billiard balls is a good example of a two-dimensional collision. If you shoot the cue ball toward a second ball, originally at rest, they bounce off each other. But if the cue ball misses the dead center of the second ball, they will split apart and will not continue to move in a line after the collision.

c. Because the initial momentum of the system was all in the x-direction, the final momentum of the system must also be *only* in the x-direction. This means that the y-components of the momenta must add to zero.

d. Both of the following scenarios are possible for a two-dimensional collision:

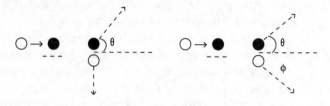

e. The scenario below is *not* possible for a two-dimensional collision. Notice after the collision, there is a *y*-component to the momentum, where there was initially no *y*-component.

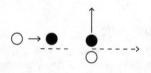

Whenever you have a two-dimensional collision, use conservation of momentum separately in both the horizontal and vertical direction.

D. IMPULSE

1. If a force is applied to an object for a certain amount of time, the object's velocity, and therefore momentum, will change. Impulse is equal to the net external force on an object times the time interval over which that force is applied.

$$\text{Impulse} = F \times \Delta t$$

2. A change in momentum is also known as an impulse.

$$\text{Impulse} = \Delta(m \cdot v)$$

3. Impulse can be measured in units of Newton Seconds (N·s).

4. Remember Newton's Third Law states that every action has an equal and opposite reaction, which means that if one object applies a force on a second object, the second object applies an equal and opposite force to the first object.

5. Newton's Second Law was actually originally written in the form of impulse and change in momentum. It is equivalent to the equation, $\sum F = ma$.

$$\sum F_{external} = ma$$

$$\sum F_{external} = m\frac{\Delta v}{\Delta t}$$

$$\sum F_{external} \times \Delta t = m\Delta v$$

$$\sum F_{external} \times \Delta t = \Delta p$$

6. When analyzing a situation with a collision, based on the given information, determine whether to apply the impulse momentum theorem or the conservation of momentum. For example, where two balls collide together and the mass and initial velocities of the balls are known, the situation will most likely be analyzed using conservation of momentum. If a ball is bounced against a wall and you are asked about the force given to the wall or the ball, the situation will most likely be analyzed using the impulse momentum theorem.

7. Automobile companies use bumpers, crumple zones, and airbags, as well as the impulse momentum theorem to reduce the force on passengers during a collision. If a car traveling at 60 mph is going to come to rest after a collision, there is nothing that can be done to alter the change in momentum of the passengers. However, remembering that the change in momentum is equal to the impulse, it is possible to change the time over which the collision takes place, and hence change the force on the passengers. Bumpers, crumple zones, and airbags increase the time over which the passenger's momentum is changed. This increase in time decreases the amount of force applied to the passengers and saves lives.

$$\Delta p = \text{Force} \times \text{Time}$$

8. Compare this to the technique of catching a water balloon. A water balloon thrown at a wall will break because the time of impact is very small, meaning that the force imparted to the balloon is very large. It is possible to successfully catch a water balloon if you cradle the balloon and pull it back toward your body while you slow it down. The balloon is coming to rest for a much longer period of time, so a much smaller force is applied, allowing the balloon to remain intact.

9. Graphs of force vs. time.

    a. Graphs of force vs. time provide an easy way to calculate impulse. The area under a force vs. time graph will be equal to the impulse given to the object.

    b. If the force is constant, the area under the curve is easy to calculate.

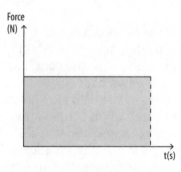

    c. If the force is constantly changing, the area is still easy to calculate. Simply find the area under the triangle ($\frac{1}{2}$ base × height).

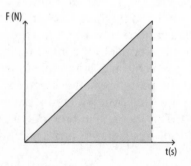

d. In real experiments, the force vs. time graph will not be linear, but will change over time. This force vs. time graph could represent a collision. Where the force first begins to increase is where the collision begins and where the force returns to zero is when the collision is over.

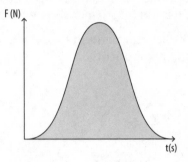

*If asked to calculate the area under a curve as shown here, estimate the area from the numbers given on the axis. Don't worry about getting the **exact** impulse. The exact answer won't matter—you'll either be asked to estimate the area on a multiple-choice question or asked to explain how you would calculate the impulse on a free-response question.*

E. PRACTICE MOMENTUM AND IMPULSE QUESTIONS

1. A ball traveling at speed *v* impacts a wall and rebounds at speed *v* as shown. What is the change in momentum of the ball because of the collision?

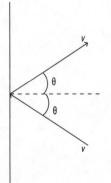

a. You may be tempted to say that because the speed of the ball doesn't change, there is no change in momentum. But, remember, momentum is a vector, and because it changes direction, there *is* a change in the momentum. Even if the ball was traveling straight toward the wall and rebounds in the opposite direction at the same speed, there is still an impulse and hence a change in momentum.

b. Break each momentum (initial and final) into horizontal and vertical components.

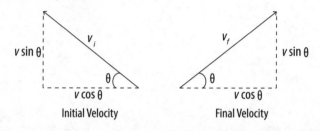

Initial Velocity   Final Velocity

c. Compare the initial and final components. Notice the vertical components (initial and final) have the same magnitude and are in the same direction. This means that there is no change in momentum in the vertical direction. [This should also make a lot of sense conceptually. Remember if there is no change in momentum, there is no impulse, and hence no force acting on the ball vertically from the wall.]

d. In the horizontal direction, the momentum goes from $mv\cos\theta$ left to $mv\cos\theta$ right. The change in momentum is:

$$\Delta p = p_f - p_i$$

$$\Delta p = mv\cos\theta - (-mv\cos\theta) = 2mv\cos\theta$$

The change in momentum (equal to the impulse) is 2mvcosθ and is directed perpendicularly away from the wall.

$$\Delta p = 2\,mv\cos\theta$$

F. CENTER OF MASS

1. The center of mass of an object is the point on an object at which all the mass of the object is distributed equally around.

2. Another way to think about center of mass is through balance. The center of mass is the point where an object would naturally balance because the distribution of the mass is even on all sides of it. If you were to balance a meterstick on your finger, it would balance right in the center, on the 50 cm mark, because it is a uniform object. If you tried to balance a spoon on your finger, it would balance closer to the head of the spoon, due to the greater distribution of mass on that side.

   a. The center of mass of a long uniform rod will be right in the middle.

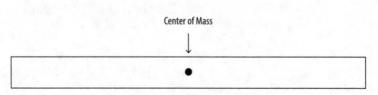

Center of Mass

   b. If that rod is weighted more on one side than the other, like a baseball bat, the center of mass will be shifted to the right of center, more toward the heavier end.

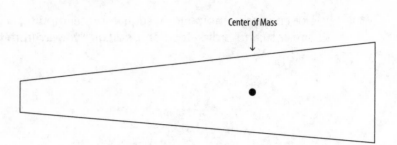

Center of Mass

3. You can also find the center of mass of several discrete objects.

   a. For example, the center of mass of three boxes of mass *m* connected by a very light (i.e., massless) rod would be right in the middle of the boxes.

   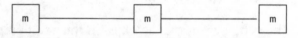

   b. However, if the boxes had different masses (as shown below), the center of mass again would be shifted to the right of center, toward the heavier end.

   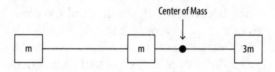

   Center of Mass

4. To find the center of mass of a two-dimensional object, think through the solution.

   a. For example, if four identical boxes are organized at the corners of a square, the center of mass would be at the center.

   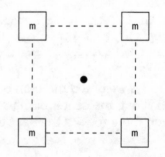

b. If one of the boxes is removed, the center of mass will be shifted away from the center and down toward the bottom left.

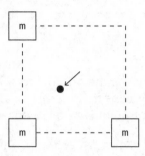

Test Tip

Using logic is usually enough for an AP Physics 1 exam question related to center of mass. If given masses and lengths and the answers are numerical, you will probably be able to figure out where the center of mass is without using an equation. Use symmetry, as discussed above.

5. If given a problem where the center of mass has to be calculated, the equation for finding the center of mass is:

$$x_{cm} = \frac{x_1 m_1 + x_2 m_2 + x_3 m_3 + \ldots}{m_1 + m_2 + m_3}$$

6. *X* is the distance from each mass to some chosen reference point, and *m* is the mass of each piece. Note you are solving for a location, a point in space. Also notice this equation not only considers the mass, but the distribution of the mass as well. This equation can also be applied to the vertical direction, so the exact point of the center of mass of the object as (x_{cm}, y_{cm}) can be determined.

7. To use this equation, first choose a reference point from which to measure the center of mass. Consider the example that follows. If looking for the center of mass of the system of the two 5 kg spheres,

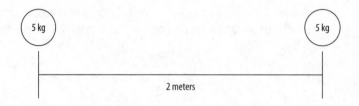

2 meters

we could choose the right-hand sphere to be our reference point. This means the distance, x, of the rightmost mass from the reference point is zero. So our equation will look like this:

$$\frac{(0)(5\ \text{kg}) + (2\ \text{m})(5\ \text{kg})}{(5\ \text{kg} + 5\ \text{kg})} = \frac{10}{10} = 1\ \text{meter}$$

8. This means that the center of mass of the system is one meter in the positive direction from the right-hand sphere, which would put it in the middle of the system, right where we expect it.

Test Tip

There are many potential reasons why you might need to know where the location of the center of mass of an object or system is. On the multiple-choice section of the exam, you might be asked to find the center of mass of an object or of a system. More likely, you will be asked about the location of the center of mass because you'll need to understand what your object or system is going to do.

For example, when we were reviewing linear motion, we were treating objects as if the whole mass of the object was located at the center of mass. In the case of a rocket being shot off a cliff, it is the center of mass of the rocket that makes a parabola. What happens if the rocket happens to be a firecracker that explodes into a million pieces at some point during the parabola? Well, although it would be almost impossible to track the motion of every single piece, if we did, the center of mass of the exploding firecracker's trajectory would still be a parabola.

9. When two people push off each other on ice, the only force would be their mutual force, so there would be no net external force. This means that the center of mass of the system *can't move.* As the two people slide apart from each other, they slide at a rate that keeps the center of mass at rest.

10. Two people are standing on the ice holding a rope to pull themselves together. There are no net external forces on the system, which includes the two people and the rope. Again, the center of mass *can't move.* So where do they meet up? Why, at the center of mass, of course!

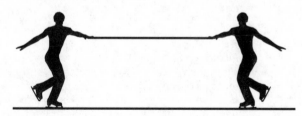

Rotation

A. ROTATIONAL KINEMATICS

1. Understanding "new" variables.

   a. When thinking about rotational kinematics, think about what you already know. Many quantities in rotational motion are analogous to those used in translational motion.

   b. Angular displacement is a measurement of the number of radians that something has rotated through. This quantity is represented by the symbol θ (theta) and is measured in radians. For example, a child riding on a carousel will have an angular displacement of π radians after the child has ridden the carousel halfway around. (This is analogous to linear displacement that is measured in meters.)

   c. Angular velocity is a measurement of an object's rotational rate. This quantity is measured in radians per second (1/s) and is represented by the symbol ω (omega). (This is analogous to linear velocity that is measured in meters/second.)

Test Tip

Remember that angles are dimensionless. When measuring an angle, you are measuring the ratio of the arc length to the radius, which doesn't need a unit.

This is important when multiplying the angle by another number. For example in $X = \theta R$, the angle in radians times the radius in meters is equal to a distance in meters, NOT radians × meters.

d. Angular acceleration is a measurement of an object's rate of change of angular velocity (the speeding up or slowing down of the rate of rotation). It is measured in radians per square second $(1/s^2)$ and is represented by the symbol α (alpha). (Angular acceleration is analogous to the linear acceleration a, which is measured in m/s^2.) It is different from centripetal acceleration (a_c) (discussed in Chapter 6), which is the rate at which the direction of an object undergoing uniform circular motion is changing.

e. At times, you will need to convert between rotational and translational quantities. To do this, remember one full revolution is equal to 2π radians. For example, if a penny on a record player has rotated through 4π radians, what distance has the penny traveled during that time? Use the following equations to convert angular position, velocity, and acceleration to linear quantities:

$$x = \theta R$$
$$v = \omega R$$
$$a_{\text{tangential}} = \alpha R$$

The linear displacement (x) represents the distance traveled by one particle on the object. It is related to the angular displacement by the equation:

$$x = \theta R$$

An easy way to visualize this is to think of a bicycle wheel that has made one full revolution, which is 2π radians. Multiplying this angular displacement by R gives us $2\pi R$, or the circumference of the wheel, which is the total distance a particle on the edge of the wheel has traveled. The linear velocity and acceleration are the instantaneous tangential velocity and instantaneous tangential acceleration of the object at that point in the object's rotation, the direction of which will be tangent to the path at that point. The tangential acceleration is different and separate from the centripetal acceleration discussed in Chapter 6. The centripetal acceleration represents the rate of change of direction of an object undergoing circular motion, while the tangential acceleration represents the rate of change of speed (speeding up or slowing down) of a rotating object. It is possible for an object to have centripetal and tangential acceleration, if it is rotating and changing speed.

f. The big three kinematic equations of motion used to discuss linear motion can still be used with rotational quantities by replacing angular quantities for linear quantities.

$x_f = x_o + v_o t + \dfrac{1}{2}at^2$	$\theta_f = \theta_o + \omega_o t + \dfrac{1}{2}\alpha t^2$
$v_f = v_o + at$	$\omega_f = \omega_o + \alpha t$
$v_f^2 = v_o^2 + 2a(\Delta x)$	$\omega_f^2 = \omega_o^2 + 2\alpha(\Delta\theta)$

Test Tip

Solving a problem on the AP Physics 1 exam using the kinematic equations is unlikely for the same reason you won't be asked to solve a problem about a rock falling off a cliff— it's too basic and it doesn't test your knowledge of physics. You might, however, be asked to solve for a variable based on some known conditions or to rank different situations based on angular acceleration or angular velocity. In either situation, you will need to have a conceptual understanding of what the kinematic equations mean.

B. ROTATIONAL INERTIA (MOMENT OF INERTIA)

1. There's a rotational analogy for position, velocity, and acceleration, but what about mass? Does mass (or inertia) matter when an object is rotating? The answer is yes, but it's more than just mass that matters, it's also how the mass is distributed.

2. While inertia is a measurement of an object's resistance to change in its linear motion, an object's rotational inertia refers to an object's ability to resist changes in its rotational motion. An object with a large rotational inertia will be difficult to get to rotate (or stop rotating), and an object with a smaller rotational inertia will be easier to start or stop.

For example, try to spin your pencil in your hand just holding the end of it. Then switch to the middle and give it a spin. It's easier to rotate when you're holding the middle of the pencil, not just because the mass is different, but because the mass is distributed differently around the axis of rotation.

3. The rotational inertia of an object is affected by the mass of the object and how far away from the rotation point the mass is. The further the mass is from the rotation point, the bigger the rotational inertia of the object.

4. The rotational inertia of an object or a system is represented by the symbol I and is measured in $kg \cdot m^2$.

You will probably not be asked to calculate the rotational inertia of objects on the AP Physics 1 exam. If you need the rotational inertia of an object for a calculation, it will be given to you. You may, however, be asked to rank objects based on their rotational inertias.

For example, a solid cylinder and a hoop with the same mass and same radius will have different values of rotational inertia because the mass is distributed differently. Which one will have the smallest rotational inertia as it is rotated around its middle? The object that has more mass closest to the rotation point will have the smallest rotational inertia, so in this case, it would be the solid cylinder, which has more of its mass closest to the middle.

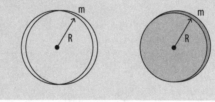

5. The equations for the rotational inertia for several common shapes are seen below for your reference. There is *no* need to memorize these equations.

Discrete Objects	$I = \sum mR^2$
Point Mass (*M*) Moving in a Circle of Radius R	$I = MR^2$
Hoop	$I = MR^2$

Discrete Objects	$I = \sum mR^2$
Solid Cylinder/Disk 	$I = \dfrac{1}{2}MR^2$ Notice the height or length of the cylinder does not matter; it is the same as for a flat disk.
Long Rod about the Center 	$I = \dfrac{1}{12}ML^2$
Long Rod about the End 	$I = \dfrac{1}{3}ML^2$
Hollow Sphere 	$I = \dfrac{2}{3}MR^2$
Solid Sphere 	$I = \dfrac{2}{5}MR^2$

6. If you are given a system of rotating objects and you need to find the rotational inertia of the system, use the equation: $I = \sum MR^2$, where M is the mass of each object, and R is the distance of each mass from the axis of rotation. To use this equation, multiply each mass by how far it is from the rotation point squared and then add up the values for each mass.

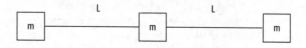

a. For example, three masses of mass m, attached to massless rods of length L, rotated around the left mass, would have a rotational inertia of:

$$\sum MR^2 = M(0^2) + M(\ell^2) + M(2\ell)^2 = 5M\ell^2$$

b. If you rotate this same system around the middle, it will have a rotational inertia of:

$$\sum MR^2 = M(\ell^2) + M(0^2) + M(\ell)^2 = 2M\ell^2$$

Notice, although the same object is rotating, the moment of inertia is less than when it was rotated around one of the ends because more of the mass is now closer to the axis of rotation.

C. TORQUE

1. Torque is the rotational analog to force. The symbol for torque is τ (tau) and is measured in Newtons times meters ($N \cdot m$). While a net force acting on an object will cause a linear acceleration, a net torque acting on a rotating object will result in an angular acceleration.

Torque is the product of a force and a distance, just like work, and has the same units. Torque is NOT WORK.

Torque is a cross product, is a vector quantity, and has a direction that is either positive (counterclockwise) or negative (clockwise). Torque has a sinθ term that puts a constraint on the quantity, requiring the force (or at least a component of the force) applied to be perpendicular to the lever arm. Think of it this way: if you push along the lever arm of an object that is free to rotate, it won't spin; however, if you push perpendicular to the lever arm, it will spin!

Work, however, is a dot product, a scalar quantity, and has no direction. Work has a cosθ term that also puts a constraint on the quantity, only this requires the force (or at least a component of the force) applied to be parallel to the displacement.

2. The torque produced depends not only on the force that is applied, but also on where that force is applied. $\tau = \vec{d}\vec{F}\sin\theta$, where $\vec{d}$ (the lever arm) is the distance between the rotation point and the point where the force is applied. $\vec{F}$ is the force, and θ is the angle between the $\vec{d}$ and $\vec{F}$ vectors.

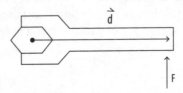

a. Another way to write the torque equation is $\tau = F\perp d$, where $F\perp$ is the component of the force perpendicular to the lever arm. For example, if you push on the handle of a door at an angle, whether you use $\tau = \vec{d}\vec{F}\sin\theta$ or $\tau\, F\perp d$, either way you end up with the same torque value.

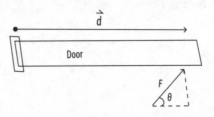

3. Newton's Second Law also has a rotational analog. In the same way that linear acceleration is caused by a net force, angular acceleration is caused by a net torque.

   a. Newton's Second Law for rotation can be stated as:

$$\sum \tau = I\alpha$$

Test Tip *On the AP Physics 1 exam, you might be asked to draw a force diagram. (Do not draw a free-body diagram, since it matters where the force is applied.) Drawing a force diagram is almost always a good idea when analyzing forces and torques.*

4. Rotational Statics

   a. There are two types of equilibrium: static and dynamic. An object that is in static equilibrium will have a net torque and a net force of zero. In situations involving dynamic equilibrium, the linear and rotational accelerations are zero (meaning the linear and rotational velocities are constant). While the linear and rotational acceleration of a system in static equilibrium are also zero, it specifically means no motion (static), so linear and rotational velocity are zero.

   b. Analyzing a situation where an extended object is in static equilibrium is common. Look at the diagram below showing a flagpole that extends horizontally off the side of a building. It is kept horizontal by a wire attached to the building. The mass of the flagpole is m, its length is L, and the angle between the wire and the flagpole is θ. In such cases, consider the sum of the forces and the torques are zero.

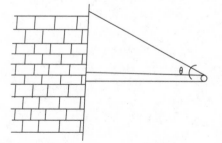
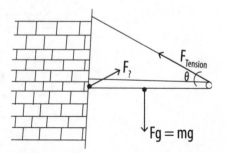

c. When drawing the force diagram, remember that the weight of the flagpole acts as if it was at the center of mass (this is a uniform pole, so the center of mass is in the center), and the tension acts at the end where the cord supports the pole. Are there any more forces? What about a force from the wall? How can we tell if there is one, and in what direction it acts?

d. Keep in mind that if an object is in equilibrium, the sum of the forces is zero. Looking at the two forces already drawn, there is an upward and a downward vertical component. It is possible that these two components are equal to each other and will cancel out. It may not be necessary to have a vertical component of force from the wall. But what about a horizontal component? So far we only have one horizontal component from a piece of the Tension. If this flagpole is supposed to be in static equilibrium, it must at least have a horizontal component of force acting on it from the wall to make a net x directional force of zero.

e. Now that you've drawn all the forces, pick a pivot point and find the torques.

f. But if this object isn't rotating, where is the pivot point? You get to choose the pivot point wherever you would like. This is similar to choosing a coordinate system with a kinematics analysis—there isn't a wrong way to do it, but if you choose wisely, it makes the problem much simpler to analyze. With that in mind, think of what happens in terms of the torque for any forces acting at the chosen pivot point. Since $\vec{d}$ is zero, the force does not create a torque. If there are forces that don't have known values, choose the point where one of those forces acts for your pivot, because it means you have fewer torques to analyze.

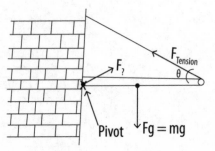

g. In this example, a good place for the pivot would be where the flagpole is hinged to the wall. This means that the force that we know very little about will create no torque.

h. Now that you know where the pivot will be and the forces that are acting on the flagpole, sum the torques. Remember because the flagpole is in static equilibrium, the sum of the torques will be equal to zero.

$$\sum \tau = 0$$

$$\tau_{mg} - \tau_{T} = 0$$

$$(mg)\left(\frac{L}{2}\right)\sin(90) - (F_{T})(L)\sin(\theta) = 0$$

If given the mass, the length of the flagpole, and the angle at which the wire is attached to the pole, calculate the force of tension in the wire.

Test Tip

Remember it is unlikely that you will have to solve a complicated mathematical equation on the AP Physics 1 exam, but you may have to explain how you would solve, or be asked if you have enough information to solve. The only way you'll know the answer to these questions is to start with a force diagram.

5. Rotational Dynamics

a. When the net torque on an object is *not* zero, the sum of the torques will cause an angular acceleration.

$$\sum \tau = I\alpha$$

b. Approach this type of problem the same way you would approach a situation with unbalanced *forces*. Draw a force diagram (or several force diagrams), pick a pivot, and think through the sum of the torques created by the forces.

6. Rolling

    a. A special case of rotational dynamics is the case of a rolling object. Even though it seems very different from the dynamics cases you've seen, follow the same steps as in other dynamics cases.

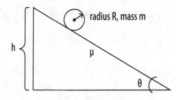

radius R, mass m

h

μ

θ

    b. A hoop of mass *m* and radius *R*, rolls down a steep incline of height *h* that makes an angle θ with the horizontal. Here are some questions that could be asked:

      i. Calculate the acceleration or find the speed of the hoop at the bottom of the incline.

        • Solve for the acceleration using forces and torques. Draw a force diagram and choose a pivot.

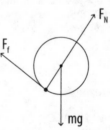

F_N

F_f

mg

        • There will be three forces acting on the hoop—the gravitational force, the normal force, and the frictional force.

        • If you choose the pivot to be at the center of the hoop, only one force will create a torque. The gravitational force gives no torque because the force acts at the pivot we chose. The normal force also gives no torque because the angle between *d* and the normal force is 180°.

- Therefore, there is only one force, friction, when summing the torques.

$$\sum \tau = I\alpha$$

$$(F_f)R\sin 90 = (MR^2)\left(\frac{a}{R}\right)$$

$$F_f = Ma$$

- Remember that even though we're using torque here, it is also necessary to sum the forces as well! Once you sum the forces, you'll put the equation $F_f = Ma$ into the sum of the forces equation.

- If asked for velocity at the bottom of the ramp, use a kinematic equation (or the ideas involved in a kinematic equation) to find the velocity after finding the acceleration of the hoop. Another approach to finding the velocity would be to use a conservation of energy analysis. The initial energy at the top of the ramp would be gravitational potential energy, which gets converted into kinetic energy ($\frac{1}{2}mv^2$) and rotational energy ($\frac{1}{2}I\omega^2$, as discussed below). You can then convert ω to v using the relationship $v = \omega R$ and solve for v.

c. More rotational questions.

  i. Explain why the hoop will roll instead of slide. (Hint: if there's not enough friction, it will slide.)

  ii. Calculate the minimum coefficient of friction, so the hoop will roll instead of slide. (Hint: sum the torques and forces and see what you can figure out.)

  iii. If the hoop were replaced with a disk with the same mass and radius, how would the acceleration and velocity change? (Hint: how does the rotational inertia affect the acceleration of the hoop?)

  iv. Which one (the hoop or the disk) would get to the bottom first? (Hint: how does the acceleration of the hoop affect the time it takes to get to the bottom? How does the size of the rotational inertia affect how long it takes something to speed up or slow down?)

Test Tip

Remember the relationships x = θR, v = ωR, and a = αR only apply if the object is rolling without sliding. If there is any sliding mixed with the rolling, these relationships do not apply.

D. ROTATIONAL KINETIC ENERGY

1. Rotational kinetic energy is analogous to kinetic energy. The equation for rotational kinetic energy is $KE_{rot} = \frac{1}{2}I\omega^2$, where I is the rotational inertia and ω is the angular velocity. This works just like translational kinetic energy—if an object is rotating, it will have kinetic energy—and it is measured in Joules (J).

2. The work-kinetic energy theorem can be used on an object that is rotating or will begin to rotate. If there is a non-conservative force acting on an object or a system, the work done by that force will equal the change in the mechanical energy of the system. Depending on the situation, this work can change the kinetic energy to linear and/or rotational.

3. **Example:** A pulley (with radius R and mass m) with a mass $2m$ is hanging from a long string. Use energy to find out how fast the hanging mass will be traveling after falling a distance d if it was released from rest.

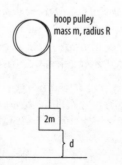

hoop pulley
mass m, radius R

2m

d

a. Since the only external force on the system (the pulley, the string, and the mass) is gravity, there are no non-conservative forces. Mechanical energy is conserved.

$$E_o = E_f$$

b. At the beginning of the problem, there is only gravitational potential energy. At the end of the problem, the total mechanical energy consists of linear kinetic energy and rotational kinetic energy.

$$(2m)gh = \frac{1}{2}(2m)v^2 + \frac{1}{2}I\omega^2$$

c. From here, depending on what other information is given, you are almost ready to solve for the velocity of the block. The rotational velocity of the pulley will be related to the linear speed of the block by the equation

$$v = \omega R$$

Knowing the rotational inertia (I) of the pulley, substitute in and solve for v.

d. If the pulley was a simple hoop (remember the rotational inertia of a hoop is $I = mR^2$), the energy equation becomes:

$$(2m)gh = \frac{1}{2}(2m)v^2 + \frac{1}{2}(mR^2)\left(\frac{v^2}{R^2}\right)$$

Notice the radius of the pulley and the m's will cancel out, leaving:

$$2gh = \frac{3}{2}v^2$$

$$v_{final} = \sqrt{\frac{4}{3}gh}$$

Note: This pulley question can be solved by summing torques and forces. Give it a try and see if you get the same answer!

Test Tip — *Although it is unlikely that you will be asked to solve for the speed of the block, you may be asked if the mass of the pulley was twice as large, what would happen to the speed of the block? Or, if the radius of the pulley were half as large, what would happen to the acceleration of the mass? Be sure you understand that energy is conserved and how to work your way through to the answer.*

Test Tip — *When you check a symbolic solution, be certain that you check the units and that your answers make sense.*

4. Rotational vs. Translational Energy

   a. It is important to know when you will be using translational energy, when you'll be using rotational energy, and when you would use both.

   b. For example, a bicycle wheel, sliding along a frictionless table, only has translational kinetic energy, since it is sliding and not rotating.

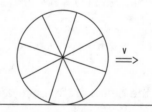

   c. If you had the same bicycle wheel suspended from the ceiling, and the wheel was rotating freely, you would use rotational energy.

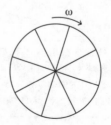

d. If the bicycle wheel is rolling down the street, it has rotational *and* translational energy.

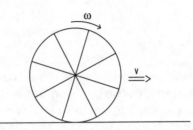

E. ANGULAR MOMENTUM

1. Angular momentum is the rotational analog of linear momentum. It can be calculated using the equation: $L = I\omega$, or $L = mvR \sin \theta$ and is measured in kg $\cdot$ m^2/s.

2. Angular momentum is conserved if there is no *net external torque,* just as linear momentum is conserved if there is no external force.

Test Tip

Angular momentum for an object rotating on its own axis can be calculated using $L = I\omega$. If an object is rotating around an external point, use $L = mvR \sin \theta$, where mv is the momentum of the object, R is the distance between the pivot point and the object, and θ is the angle between that R vector and the momentum vector.

3. If an ice skater spins with her arms stretched out and then pulls her arms in, you expect that her angular velocity will increase (she will spin faster), but why? Initially, with her arms stretched out, she has a large rotational inertia, and when she pulls her arms in, she has a smaller rotational inertia. There is no net external torque on her while she is pulling her arms in, since the force she is applying to bring her arms in is internal to the system. This means that her angular momentum is conserved,

so her initial angular momentum and her final angular momentum are the same:

$$L_i = L_f$$
$$I_i \omega_i = I_f \omega_f$$

If the final rotational inertia is less than the initial rotational inertia, it means the final angular velocity must be larger than the initial angular velocity. Even though her angular momentum remains constant as she pulls her arms in, her rotational kinetic energy increases! She must be doing work on her arms, which added to the rotational kinetic energy.

4. The rotation of the planets around the sun is another example where the conservation of momentum can be observed. As the planets orbit, they orbit in an ellipse with the sun at one focus. This means that the planet is closer to the sun at some points in its orbit. The only force that is acting on the planet as it orbits is the force of gravity between the sun and the planet. That force is a centripetal force, so it exerts no torque. (It cannot exert torque because the angle between the force and the lever arm (d) is 180°, making the torque = zero.)

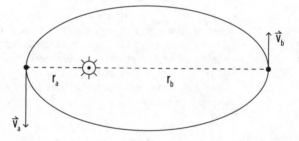

a. Because the net torque on the planet as it travels is zero, the angular momentum is constant throughout the orbit.

$$L_a = L_b$$
$$mv_a r_a = mv_b r_b$$

b. Through conservation of angular momentum, when the planet is furthest from the sun (when r is large), the speed of the planet must be small, and when the planet is close to the sun (when r is small), the speed of the planet must be larger. This result is also known as Kepler's Third Law. Kepler discovered, by analyzing data, that planets moved faster when they were closer to the sun, but he didn't have a good explanation as to why.

F. COLLISIONS

1. Angular momentum is conserved in a collision if there is no external torque on the system during the collision.

2. It is possible that in a collision, you will need to use conservation of angular momentum *and* linear momentum, and it will be important for you to understand when and why both are conserved.

3. A mass of clay in empty space travels toward a long pipe at rest. What will happen directly after the collision if the clay hits and sticks to the center of mass of the pipe?

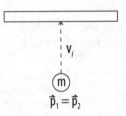

a. During this collision, there are no external forces (since we are in empty space, assume that the influence of other forces, such as friction, is negligible) and linear momentum will be conserved.

b. Since the clay is going to hit at the center of mass, the clay will exert no torque on the pipe and will cause no rotation to occur.

c. In this case, use conservation of linear momentum to calculate the linear speed of the clay-pipe system.

4. What if the clay hits the end of the pipe instead of in the middle?

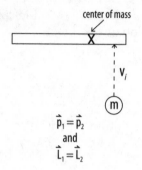

a. There still are no external forces, so linear momentum is still conserved.

b. There is still also no net torque on the system. However, the collision *will* cause the clay-pipe system to rotate around the center of mass of the *system.*

c. In this case, use conservation of linear and angular momentum.

5. What if the pipe were secured to something on the end so that it was free to rotate, but not to move linearly?

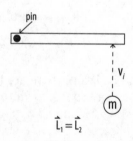

a. Now there is an external force, from the pivot, so linear momentum will not be conserved in this case.

b. There is no net torque on the system. (The pivot does not give a torque because it acts *at* the rotation point.) The angular momentum is conserved in this collision.

c. In this case, use conservation of angular momentum.

The rules used for conservation of kinetic energy during linear momentum apply here. In an elastic collision, the total kinetic energy is conserved. (This can be rotational and/or translational kinetic energy.) In an inelastic collision, there will be a loss of kinetic energy during the collision. Inelastic collisions are more common.

Simple Harmonic Motion

 A. **HOOKE'S LAW**

1. Hooke's Law describes the force that a spring will apply when stretched or compressed from its equilibrium position.

$$F_s = -kx$$

Where F_s is the force applied by the spring, x is the displacement from equilibrium, and k is the spring constant of the spring. The spring constant is measured in Newtons per meter (N/m) and every spring has its own unique spring constant.

2. The negative sign in Hooke's Law shows the force that the spring applies is opposite to the direction of the stretch. For example, if you pull a spring to the right past equilibrium, the force applied by the spring will be toward the left (toward equilibrium).

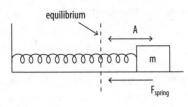

3. Hooke's Law shows the spring force is a *restoring force*. This means that the force applied by a spring will always pull a displaced object back toward equilibrium.

A question on the AP Physics 1 exam involving Hooke's Law might ask you to predict what would change if a spring in a problem was replaced with a different spring with a larger spring constant. You wouldn't necessarily need to calculate the answer, but you will need a strong understanding of what Hooke's Law means.

4. The equilibrium point for a mass on a spring is where the net force is zero.

5. Notice the spring force (F_s) for an ideal spring is directly proportional to the displacement from equilibrium, so at equilibrium, the spring exerts zero force, and as the displacement from equilibrium increases, so does the force. This means the force is not a constant and changes, depending on how much the spring has been stretched or compressed. Finding the work done on a spring cannot be calculated as $W = F \cdot d$, as this is only true of constant forces. There are, however, two ways to find the work done on the spring. One way is to calculate the total energy stored in the spring $U_s = \dfrac{1}{2}kx^2$, meaning the total work done on the spring is stored as the elastic potential in the spring. Another way is to find the area under the curve in a force vs. displacement graph for a spring, as outlined below.

6. Because the spring force for an ideal spring is directly proportional to the displacement, a graph of the force of the spring vs. the displacement would be linear, and its slope would be the spring constant of the spring. If asked to discuss how to check to see if a certain spring behaves as an ideal spring, measure the restoring force as a function of the stretch of the spring, and if the graph of F_s vs. x is linear, then you have an ideal spring.

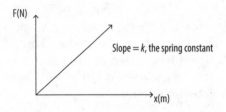

Slope = k, the spring constant

Remember that the area under a force vs. displacement graph is the work done by that force (see Chapter 7). The AP Physics 1 exam is going to link topics wherever possible. If you are given a graph of the spring force vs. displacement, remember the spring constant will be the slope and the area under the curve will be the work done by the spring.

B. SIMPLE HARMONIC MOTION

1. Imagine the same mass on the spring, displaced to the right of equilibrium. If the mass sits on a frictionless surface, what will happen to it when released from rest?

   a. The restoring force will pull the mass back toward the equilibrium point, accelerating the mass from rest. It will not stop at the equilibrium point because even though the force there is zero, the mass will have momentum, which will cause the mass to continue past the equilibrium point.

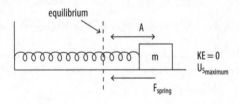

b. At the equilibrium point, the spring force on the mass is zero (because the spring is neither compressed nor stretched), so there is no potential energy stored in the spring. All the energy of the system at this point is kinetic energy. The velocity of the mass is at a maximum as the mass passes the equilibrium point.

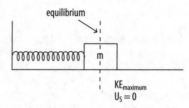

c. After the mass passes the equilibrium position, it begins to compress the spring. This in turn causes the force to apply a restoring force, pushing the mass to the right. This rightward force (and acceleration) causes the mass to slow, and the mass will stop momentarily at $-A$.

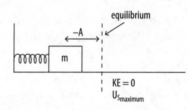

2. The restoring force will cause the mass to oscillate around the equilibrium position with a period T, measured in seconds. The period is defined as the time it takes to cycle through one full oscillation. The reciprocal of the period is the frequency $\left(f = \dfrac{1}{T} \right)$ and is the number of full oscillations per second. The frequency is represented by f and is measured in 1/s or Hertz (Hz).

3. Simple harmonic motion is defined as periodic motion that occurs because of a restoring force.

4. Two examples of simple harmonic motion are the mass oscillating on a spring and a pendulum. In the case of the mass oscillating on a spring, the restorative force is provided by the spring force. In the case of a pendulum, the restorative force is provided by the gravitational force.

Test Tip

On the AP Physics 1 exam, you will be asked to analyze problems that approximate the motion as simple harmonic motion.

C. **MASS ON A SPRING**

1. The amplitude (A) of the oscillation for a mass on a spring is the distance from the equilibrium position to the position of maximum compression or maximum stretch and is measured in meters. For example, if a small mass attached to a horizontal spring on a frictionless surface is pulled 0.05 m from its equilibrium position and is released, its amplitude, A, will be 0.05m. As there is no friction, the mass will oscillate back and forth between the points 0.05m on either side of the equilibrium position.

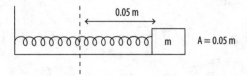

0.05 m

m   A = 0.05 m

2. The potential energy stored in the spring depends on the distance from equilibrium. The potential energy is given by:

$$U_s = \frac{1}{2}kx^2$$

Spring potential energy is measured in Joules, is a form of mechanical energy, and can be combined with gravitational and/or kinetic energy. The energy stored by the spring is largest when the displacement, x, is the largest. This means there will be the largest spring potential energy (U_s) at the maximum compression and extension and have zero spring potential energy at equilibrium.

3. The period for a mass oscillating on a spring is given by

$$T = 2\pi\sqrt{\frac{m}{k}}$$

where T is the period, measured in seconds, m is the mass attached to the spring, and k is the spring constant of the spring. Notice g, the acceleration due to gravity, and amplitude (A) are not in this equation. This implies the period for a mass oscillating on a spring would be the same on Earth or on the Moon and does not depend on the amplitude. This may seem counterintuitive, but the restorative force is provided by the spring, whose spring constant will not change if you are on Earth or on another planet. The energy is oscillating between stored elastic potential energy and kinetic energy. An object attached to the spring, if given greater amplitude, or displacement from equilibrium, results in more stored elastic potential energy and more kinetic energy at equilibrium. The object attached to the spring will be moving faster as it passes through the equilibrium position, and because x^2 varies directly with v^2, it will not take longer to cover this greater distance as it oscillates.

4. The frequency of the mass on a spring will be the inverse of the period, $f = \dfrac{1}{T}$.

Test Tip

Be careful when thinking about a mass on a spring. Since the force changes as the box slides back toward equilibrium, the acceleration of the box changes also. You cannot use any of the motion equations; they are only for motion with constant acceleration. Energy is the best approach for questions involving simple harmonic motion.

D. SIMPLE PENDULUM

1. The motion of a pendulum approximates simple harmonic motion if the angle from which it is released is small. On the AP Physics exam, you will be asked to analyze problems where simple harmonic motion can be approximated.

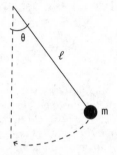

2. Again, the period does not depend on the amplitude. In this case, we have a transfer between gravitational potential energy and kinetic energy. With a greater amplitude or initial displacement, there will be a greater gravitational potential energy. This means there will be a greater amount of kinetic energy, and it will be moving faster as it passes through the equilibrium position. This means it will not take longer to cover that greater distance through each oscillation resulting from the greater amplitude. Notice that the mass does not appear in this equation. The stored gravitational potential energy is converted into kinetic energy, $U_g = KE$, $mgh = \frac{1}{2}mv^2$ the mass cancels out and does not affect the speed or the period of the pendulum as it oscillates.

3. The period of oscillation of a mass on the end of a long string is given by:

$$T = 2\pi\sqrt{\frac{L}{g}}$$

where L is the length of the string, and g is the acceleration of gravity (or the gravitational field strength) at the location of the pendulum.

4. Notice the equation for the period of a pendulum is simple and does not depend on the mass attached to the end of the pendulum or on the amplitude. In this case, the restorative force is the gravitational force (or a component of it) and the period depends on the gravitational field strength, g. This means if you had a pendulum on Earth vs. that same pendulum on the Moon, it would oscillate with a different period. Notice the period will go up as g gets smaller and will go down as g gets larger. The period depends on the length of the pendulum and will get larger as L gets larger and smaller as L gets smaller.

Test Tip

The probability that you are asked to solve for the period of a pendulum on the AP Physics 1 exam is very small. You are more likely to be asked to rank some different pendulums based on their masses and lengths. For example, the following pendulums have different lengths and are released from different (but small) angles. Rank the situations based on the period of oscillation from largest to smallest.

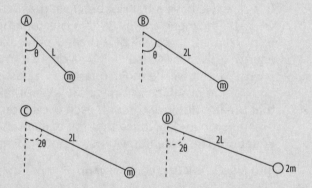

The only thing that will change the period is the length of the pendulum, so the largest period will be the pendulum with the longest string; the other variables do not matter.

The ranking should be, B=C=D > A.

5. Pendulums (and masses on springs) make good lab questions, so be prepared to graph data from a pendulum (or other simple harmonic motion) situation. If collecting and graphing data for a pendulum, collect the data for the period of the pendulum based on the length of the string. The data chart would look like this:

T (Seconds)	Length (Meters)
0.90	0.20
1.59	0.60
1.90	0.90
2.18	1.20
2.45	1.50

Once the data is collected, create a graph to represent the data. A graph of period vs. the length would not be linear, so you will need to linearize the data.

6. In order to linearize data, think about the relationship between the variables collected. In the case of the pendulum, we have the period (T) and the length of the string (L), and we know the relationship between the two is:

$$T = 2\pi\sqrt{\frac{L}{g}}$$

This can be rewritten as:

$$T = \frac{2\pi}{\sqrt{g}}\sqrt{L}$$

This makes it easier to see what needs to be graphed. Remember because the graph needs to be linear, the equation above should look like:

$$y = mx + b$$

By putting the equations on top of each other, you see what needs to be graphed.

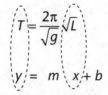

$$T = \frac{2\pi}{\sqrt{g}}\sqrt{L}$$

$$y = m \cdot x + b$$

To make the graph of your data linear for the pendulum, graph period (T) vs. the square root of the length ($\sqrt{L}$). The slope of that line will be equal to slope $= \frac{2\pi}{\sqrt{g}}$.

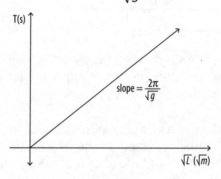

E. GRAPHS OF SHM

1. A graph of the position of an object in simple harmonic motion vs. time can be created by thinking through the motion of a mass on a spring. First, displace the mass from equilibrium, so the initial position is +x; then release, and it oscillates around zero displacement forming a sinusoidal function, which could be a sine or a cosine function. In this case, describe the position at any point in time using a cosine function because at a time of zero there is the initial maximum displacement and $\cos(0) = 1$.

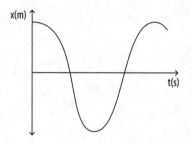

2. The equation for the position of the mass as a function of time is:

$$x = A \cos(2\pi f t)$$

where A is the amplitude of the motion, f is the frequency, and t is the time elapsed since the release of the mass.

3. The graph of the velocity of an object in simple harmonic motion vs. time can be created in a similar way. The initial velocity of the mass will be zero, just as it is released. When the mass passes back through equilibrium, the mass has the greatest velocity, and then comes back to rest again when the mass reaches the maximum compression. The velocity graph then looks like a sine graph. Also note that at a time of 0, the velocity is zero and $\sin(0) = 0$.

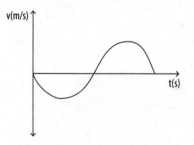

4. The equation for the velocity of the mass as a function of time is:

$$v = -A(2\pi f)\sin(2\pi f t)$$

Waves and Sound

MECHANICAL AND ELECTROMAGNETIC WAVES

> *Although only mechanical waves will be tested on the AP Physics 1 exam, you will need to understand and explain the difference between mechanical and electromagnetic waves.*

1. All waves transfer energy without permanently disturbing the medium through which they travel.

2. An electromagnetic wave is a field wave and can transmit energy through a vacuum. All waves are created by something oscillating, or vibrating, back and forth. An electromagnetic wave is created by the vibration of charged particles. Examples of electromagnetic waves include low-energy radio waves and high-energy gamma rays. The most familiar is visible light, although this comprises only a small portion of the electromagnetic spectrum.

3. A mechanical wave is not able to transmit energy through a vacuum. Mechanical waves need a medium to travel through. Sound waves can travel through air, water, or solids (such as a wall), but cannot travel through a vacuum. A mechanical wave temporarily displaces the particles of the medium from their rest position and carries energy through the medium without transporting the matter. Think of an object sitting on the surface of a lake or pond, like a bobber. As a wave passes,

the bobber will move up and down (or bob), but it will not be carried away with the wave.

4. A mechanical wave has four main properties diagrammed below:

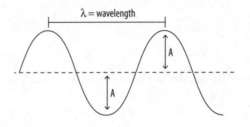

a. Amplitude, A (measured in meters), is the maximum displacement from the equilibrium position.

b. Wavelength, λ (measured in meters), is the distance between two adjacent points at the same phase on the wave, such as the distance between two crests (or the distance between two troughs), on a wave.

c. Period, T (measured in seconds), is the time required for one full wavelength or wave cycle to pass a fixed point.

d. Frequency, f (measured in 1/s, Hertz [Hz]), is the number of waves that pass a fixed point every second.

5. The frequency of a wave is inversely proportional to the period of the wave.

$$f = \frac{1}{T}$$

6. Frequency is measured in 1/s or Hz. This is a measurement of how frequently a wave cycle occurs. For example, 1 Hz means you get one wave cycle every second while 60 Hz means you get 60 wave cycles every second.

Be sure to have a clear understanding of the difference between frequency and wave speed. Imagine a person throwing baseballs at a target at 30 mph. This person throws a baseball every 10 seconds. The speed of the baseballs is 30 mph, and the frequency with which baseballs are being thrown is .1 Hz $\left(\dfrac{1 \ ball}{2 \ seconds} = 0.1 \ Hz \right)$.

Another person throws baseballs again at 30 mph, but the ball is being thrown every 2 seconds. The speed of the baseballs is still 30 mph, but the frequency at which the baseballs are being thrown is now .5 Hz $\left(\dfrac{1 \ ball}{2 \ seconds} = 0.5 \ Hz \right)$.

The baseballs are not traveling faster; there are just more of them in a given time frame. This occurs with a higher frequency wave. It does not mean the wave travels faster—it means there are more wave cycles in a given timeframe.

7. The speed of a wave is directly proportional to the frequency of the wave *and* the wavelength and is measured in m/s.

$$v = f\lambda$$

Understanding what the equation means is important. The speed of a mechanical wave depends on the medium it is traveling through. (For example, the speed of a wave on a string depends on the tension in the string, the mass, and the length of the string. If you leave the mass, length, and tension the same, the speed of the wave will remain constant.) Typically, waves travel fastest through solids, liquids, and gases, respectively. This is because a mechanical wave relies on the energy being transferred from particle to particle through the material, so a higher density will allow it to travel more quickly. Electromagnetic waves do not require a medium. They travel at the same speed through a vacuum, commonly referred to as the speed of light (although all electromagnetic waves, not just visible light, travel at this speed). The speed of light, or c, is:
c = 3.0×10^8 m/s.

On the AP Physics 1 exam, you may be asked what will happen to the speed of a wave, either on a string, or through air, or some other medium, if the frequency is doubled. Watch out! Remember the speed of a mechanical wave depends on the medium it travels through and cannot be changed by changing the frequency of the wave. If you double the frequency, the wavelength will be cut in half, so that the velocity remains constant.

8. The energy of the wave is proportional to the amplitude squared and is measured in joules.

$$E \propto A^2$$

If the amplitude of the wave is doubled, the energy carried by the wave is quadrupled.

B. TRANSVERSE AND LONGITUDINAL WAVES

1. A transverse wave displaces particles of the medium perpendicular to the direction of motion of the energy of the wave as the wave passes through the medium. For example, a pulse sent down a coiled spring by a student shaking the end up and down creates a transverse wave. In this pulse, the individual particles (or coils of the spring) travel up and down while the energy of the wave travels to the right.

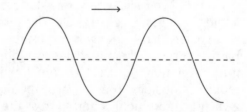

2. A longitudinal wave displaces the particles of the medium parallel to the direction of motion of the energy of the wave as the wave passes through the medium. To create a longitudinal wave on a coiled spring, gather several coils up and release them, sending a pulse down the spring. In this pulse, the individual particles (or coils of the spring) travel left and right while the energy of the wave travels to the right.

    a. Sound is the most common type of longitudinal wave. It is a pressure wave. The areas of high pressure or a high density of particles (an area of compression) is analogous to a crest on a transverse wave. The areas of low pressure or a low density of particles (an area of rarefaction) is analogous to a trough on a transverse wave. A wavelength on this type of wave is measured from one area of compression to another or one area of rarefaction to another.

C. THE PRINCIPLE OF SUPERPOSITION

1. When two wave pulses are traveling on the same string, they interfere with each other. The law that governs their interaction is called the principle of superposition. The principle of superposition states that the amplitudes of the two pulses will add as they pass through each other and come out on the other side undisturbed.

    a. Constructive interference occurs when the superposition of the waves results in larger amplitudes, as shown below. This occurs when the displacement of the pulses is in the same direction from equilibrium. If both peaks are positive or negative, the result will be a greater overall peak in that respective direction.

    b. Destructive interference occurs when the superposition of the waves results in smaller (or zero in some cases) amplitudes as shown below. This will occur when the displacement of the pulses is in the opposite direction from equilibrium. One is positive and one is negative, resulting in subtraction or cancellation.

|159

D. **WAVES REFLECT**

1. When waves bounce off a boundary, they reflect back, and the reflected wave will either be *in-phase* or *out-of-phase*.

2. In-phase reflection results in a wave with the same orientation as the original wave. This type of reflection occurs in one of two situations:

   a. For a wave on a string, an in-phase reflection occurs when the string has a loose end.

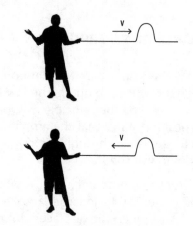

   b. For a sound wave, an in-phase reflection occurs when the sound wave is reflected from the boundary between mediums when the second medium has a lower density than the first (such as a sound wave going from water to air).

3. Out-of-phase reflection results in a wave with the opposite orientation as the original wave. This type of reflection occurs in one of two situations:

   a. For a wave on a string, an out-of-phase reflection occurs when the string has a fixed end.

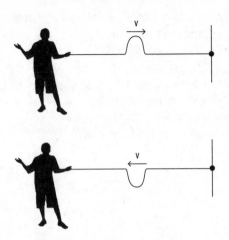

b. For a sound wave, an out-of-phase reflection occurs when the sound wave is reflected from the boundary between mediums when the second medium has a higher density than the first (such as a wave going from air to water).

TRAVELING WAVES VS. STANDING WAVES

1. A traveling wave occurs when a series of pulses are sent through a medium, and these pulses are not confined in a certain region. For example, a wave on a string is often not a traveling wave because it will quickly reach the end of the string and reflect, causing interference.

2. A standing wave is formed when a wave is confined to a space, and the pulses of the wave that are traveling through the medium are interfering with the reflected pulses, such that the wave form appears not to be moving. For example, if you send several pulses down a string, it is possible to form a standing wave on the string.

3. A standing wave can only be created by using certain frequencies called harmonics.

4. Every standing wave has points along the medium that appear to be standing still. These points are called nodes. The points of the medium that are displaced the most are called antinodes. Nodes are created at locations where destructive interference occurs—for example, where the crest of one wave meets the trough of another. Antinodes are created where constructive interference occurs—for example, where a crest from one wave meets the crest from another.

Don't confuse nodes and antinodes with crests and troughs. Nodes and antinodes describe places on a standing wave and are not part of the wave. Crests and troughs describe pieces of a traveling wave.

5. The lowest frequency standing wave that can be formed is also called the fundamental frequency or the first harmonic and is sketched below:

6. Other patterns appear as the frequency of the wave is increased. At twice the fundamental frequency (the 2nd harmonic), another standing wave pattern will be seen.

7. At three times the fundamental frequency (the 3rd harmonic), another standing wave pattern appears.

8. Guitar strings (and other stringed instruments) have a number of frequencies at which they will vibrate naturally. These frequencies are known as the natural frequencies and will depend on the tension in the string, the mass, and the length of the string.

For the AP Physics 1 exam, you will need to be able to explain the relationships between the harmonic number, frequency, wavelength, and the speed of the wave. The best way to do this is to practice sketching out the different waves and explaining how the frequencies compare to the fundamental frequency, the wavelength, and the speed. The table below is for a standing wave that has two fixed ends (such as a guitar string):

Harmonic	Length	Number of Nodes	Number of Antinodes
Fundamental (f_1) 1st harmonic	$L = \frac{1}{2}\lambda$	2	1
2nd harmonic $f_2 = 2f_1$	$L = \frac{2}{2}\lambda$	3	2
3rd harmonic $f_3 = 3f_1$	$L = \frac{3}{2}\lambda$	4	3
4th harmonic $f_4 = 4f_1$	$L = \frac{4}{2}\lambda$	5	4
5th harmonic $f_5 = 5f_1$	$L = \frac{5}{2}\lambda$	6	5
nth harmonic $f_n = nf_1$	$L = \frac{n}{2}\lambda$	$n + 1$	n

F. **RESONANCE**

1. Every object has a specific frequency that will naturally oscillate. This is known as its natural frequency. An object's natural frequency depends on the shape and material of the object. To visualize this idea, think of a swing. To get the swing to oscillate, pump your legs. If you pump your legs at the correct time, you will go higher and higher. The swing is oscillating at its natural frequency. If you pump your legs at the wrong time, you dampen the oscillation because that is not the frequency at which it would naturally oscillate. It causes destructive interference patterns. Everything has a natural frequency, even buildings and bridges.

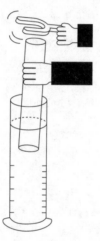

2. Resonance occurs when an object vibrating at the natural frequency of another object forces the second object to vibrate. If you strike a tuning fork and hold it over a tube partially filled with water, you won't hear the tuning fork very loudly until the natural frequency of the air in the tube matches the frequency of the tuning fork. The natural frequency of the air in the tube depends on the length of the tube above the water, and so by moving the tube up and down, you can find a length that has a natural frequency that matches the tuning fork. At this point, you will have resonance—the air in the tube vibrates at the same frequency as the tuning fork.

3. The result of resonance is a buildup of constructive interference, a large vibration, a large amount of energy, and a large sound. In some musical instruments, a mouthpiece or a reed is vibrated by the musician's mouth. When one of the frequencies of this vibration matches the resonant frequency of the air inside the instrument, the instrument will play a loud sound.

4. Each wind instrument has a set of natural frequencies that are multiples of the fundamental frequency, called harmonics, and each of these harmonics has a standing wave pattern that is associated with it. For the guitar (and other stringed instruments), the wave pattern is drawn to represent the movement of the string. In the case of wind instruments, look at the pressure in the instrument.

5. Open-Tube Instruments

   a. The simplest wind instrument is an open tube where both ends are open to the room. The pressure at both ends of the tube is fixed at atmospheric pressure. (In other words, you're not going to significantly change the pressure in the room by blowing into a flute!) Although the pressure at the ends of the tube is fixed, the pressure in the middle of the tube could be significantly larger or smaller than atmospheric pressure.

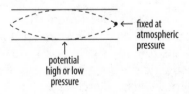

← fixed at atmospheric pressure

↑
potential high or low pressure

Test Tip

The vibration of the air in the tube is a longitudinal wave. By sketching the pressure differences as a wave, don't get tricked into thinking it is a transverse wave!

These patterns show nodes—points at which the pressure is at atmospheric pressure, at the two open ends of the tube, while the antinodes are inside the tube where the pressure can vary.

    b. The fundamental frequency (or first harmonic) in a tube which is open at both ends can be sketched with a pressure wave:

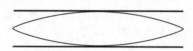

    Notice there is only a half a wavelength inside the tube, so $L = \dfrac{1}{2}\lambda$.

    c. The second harmonic has a frequency that is twice the fundamental frequency and is sketched as

    In the second harmonic, we have a full wavelength inside the tube, so $L = \lambda$.

    d. The pattern for pressure waves in open-tube instruments follows the same pattern as for stringed instruments.

  6. Closed-Tube Instruments

    a. For a closed-tube instrument, there are some interesting consequences of the pressure wave. When one of the ends of the instrument is closed, it is no longer forced to stay at atmospheric pressure, so the standing wave with the lowest frequency will be sketched as:

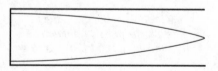

    b. Notice that there is not a half wavelength inside the tube. Here, there is only a quarter of one wavelength inside the tube. Thus, $L = \dfrac{1}{4}\lambda$.

c. For the second harmonic, the amount of nodes would double, and the sketch would look like this:

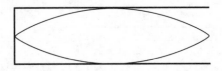

But wait: There is no pressure node at the closed end of the tube. The closed end of the tube must be an antinode. There are no even harmonics on a closed tube. They just do not exist!

d. The next harmonic will be the third harmonic (three times the fundamental frequency) and will be sketched like this:

Notice again there is an antinode at the closed end and a node at the open end. The sketch shows that for the third harmonic, $L = \dfrac{3}{4}\lambda$.

	Harmonic	Length
	Fundamental (f_1) 1st harmonic	$L = \dfrac{1}{4}\lambda$
	3nd harmonic $f_3 = 3f_1$	$L = \dfrac{3}{4}\lambda$
	5rd harmonic $f_5 = 5f_1$	$L = \dfrac{5}{4}\lambda$

(continued)

	Harmonic	Length
	7th harmonic $f_7 = 7f_1$	$L = \dfrac{7}{4}\lambda$
	9th harmonic $f_9 = 9f_1$	$L = \dfrac{9}{4}\lambda$
	nth harmonic $f_n = nf_1$	$L = \dfrac{n}{4}\lambda$

G. SOUND

1. Loudness. The loudness of sound depends on the amplitude of the sound wave, and the amplitude depends on the energy carried by the wave. Loudness is measured in decibels and is inversely proportional to the square of the distance to the sound source.

2. Beats.

   a. Sound is a pressure wave that is made up of compressions and rarefactions. (A compression is a place where the air molecules are closer together and that region has a higher pressure. A rarefaction is a place where the air molecules are further apart, and hence, that region has a lower pressure.)

   b. Sound waves can interact with each other, and their waves will interfere with each other via the principle of superposition. Where there is a region in which two compressions meet, the effect is a region of even higher pressure.

   c. Where two sound waves interfere constructively, the sound is louder, and where the two sound waves interfere destructively, the sound is softer.

   d. When two sounds of very similar frequencies are played at the same time, their waves alternately interfere constructively and destructively, which creates a pattern of loud and soft. For example, when the two waves (shown below) with very similar frequencies are played together, they combine to create one sound with alternating loud and soft patterns.

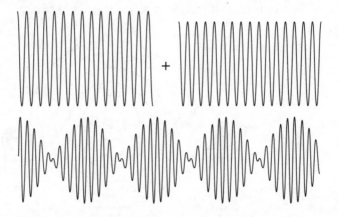

e. The beat frequency is the rate at which the volume of the sound oscillates between loud and soft and is equal to the difference in the frequencies of the notes. For example, if two tuning forks are struck, one with a frequency of 440 Hz and the other with a frequency of 442 Hz, the beat frequency is:

$$f_{beat} = |f_1 - f_2|$$

$$f_{beat} = 442 - 440 = 2 \text{ Hz}$$

This means sound will oscillate between loud and soft twice per second.

f. A piano tuner uses the phenomena of beats to accurately tune a piano. He will strike a key and a tuning fork at the same time. If the two notes are different, he will hear beats. As the piano string gets closer in frequency to the frequency of the tuning fork, the beat frequency will decrease. When the beat frequency is zero, the piano and the tuning fork have the same frequency, which means that the string will have been tuned correctly.

3. Doppler Effect

   a. Sound travels radially outward from a source in all directions. Therefore, a stationary source making a wave (sound or any other kind of wave) creates a circular (or in the case of sound and light, spherical) wave front that expands as it travels, so that it looks like this:

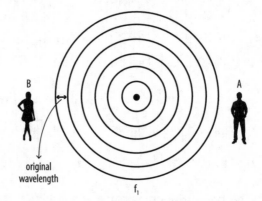

If the tuning fork has a frequency of 440 Hz, it will send out a wave front 440 times every second. The first wave that it sent, having more time to travel than the last creates the circular (or spherical) pattern.

   b. A person on either side of the tuning fork would hear the same frequency, and the frequency would match the frequency of the tuning fork.

   c. If the tuning fork is moved toward observer A (shown in the following diagram), each wave starts from a position that is closer to observer A and further from observer B. Each wave would take less time to reach observer A, causing observer A to hear a higher frequency (higher pitched) sound. (Notice also that the wavelengths seen by observer A are shorter.) Observer B hears a lower frequency (lower pitched) sound, since it takes each wave more time to reach observer B.

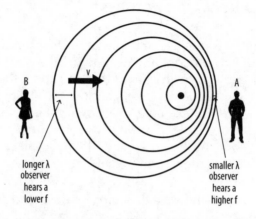

longer λ
observer
hears a
lower f

smaller λ
observer
hears a
higher f

d. The Doppler effect is the result of the relative motion between a wave source and an observer, where there is an apparent shift in the frequency of the waves. The Doppler effect does not change the actual frequency of the source, but rather the apparent frequency heard by the observer. This could be the result of a moving source or a moving observer.

e. You are most likely familiar with the Doppler effect as it relates to emergency sirens on police, fire, or EMT vehicles. If the emergency vehicle is moving toward you, the frequency of the siren seems to be higher pitched than the siren of a vehicle at rest. Once the vehicle passes you, the siren seems to be producing a sound that is lower than the pitch of the stationary vehicle.

f. Although the Doppler effect most often relates to sound waves, it actually relates to *all* waves. Astronomers use the Doppler effect to determine the speeds at which galaxies are traveling away from Earth. In fact, the understanding that our universe is expanding is based upon measurements of the Doppler shift of light coming from distant galaxies.

H. STANDING WAVES IN TUBES

1. While it is intuitive to think about the pressure wave inside of a tube, you might also need to discuss what the displacement wave looks like. Instead of thinking about the values of the pressure inside the tube, consider the ability of a molecule to move. For example, in a closed tube, a molecule at the open end is free to move back and forth, so this would be an antinode. At the closed end, the molecule is contained by the end of the tube, so this would be a node.

2. Displacement wave diagrams for open and closed tubes are shown below. Be sure you understand and can explain the pattern and the position of the nodes and antinodes.

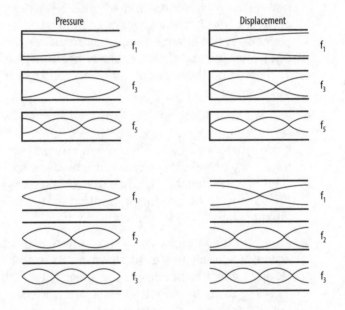

Electrostatics and Simple Circuits

A. CHARGE

1. Matter is made up of three kinds of particles: protons, neutrons, and electrons. Protons have a positive charge, electrons have a negative charge, and neutrons don't have any charge.

2. The *elementary charge* is the charge carried by a single proton or a single electron and has a magnitude equal to 1.6×10^{-19} Coulombs (C).

3. *Charge* is said to be *quantized*. This means that charge comes in distinct quantities, multiples of the fundamental unit, and does not exist over an entire gradient of values as you can't have fractions of these particles. All charges must be integer multiples of this elementary charge (you can't have half of an electron; you can have 1, 2, 100, etc.). This value for the charge on a single proton or single electron is known as the *fundamental unit of charge*, and particles, which are "singly ionized," have 1 unit of this charge.

4. All charges interact with one another; like charges repel and opposite charges attract. Thus, electrostatic forces can be attractive or repulsive, while gravitation can only be attractive. Electric charges are responsible for most microscopic forces, but can have macroscopic effects.

5. Most objects are electrically neutral because they have the same number of positive and negative charges.

6. Charge in a system, and in the universe, is always conserved.

a. If you have equal amounts of positive and negative charges in an object, the net charge will be zero, making the object neutral. The original charges still exist on the object; they don't disappear just because the object is now electrically neutral.

b. Charge can be transferred from one object to another. For example, by rubbing a balloon on your hair, you can transfer electrons, but the total amount of charge between you and the balloon stays the same. This is called the *law of conservation of electric charge*. These objects may now be attracted to one another because while one object gained electrons and has a net negative charge, the other object lost those electrons and is left with a net positive charge and the two objects attract.

7. The ability of a material to charge or to be charged depends on the ability of its electrons to be mobile. Materials that have electrons that are able to move more freely across them are called *conductors*. For example, metal is a good conductor. In conducting materials, all the excess charge resides on the surface. Materials where the electrons are less able to move do not allow charge to flow across them and are called *insulators*. Wood and rubber are good insulators.

8. Charge can be transferred between objects through friction, induction, or conduction.

a. Charging by friction. The easiest way to charge objects is charging by friction. Also called "charging by rubbing," you charge by friction when two materials pass over each other, and electrons from the material with the weakest bonds (the lower electronegativity) are ripped off and transferred to the other material (with the higher electronegativity). For example, when you rub your feet on a rug, you are charging by friction.

b. Charging by induction. If a charged object is brought close to a neutral object, the charges in the neutral object can be polarized, meaning that the positive and negative charges have been separated from each other in the object. This polarization is temporary, as long as both objects remain insulated from the ground and each other.

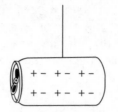

a neutral soda can hangs from
an insulating string

i. *Polarization of charge.* If you bring a negatively charged rod near an empty soda can hung from an insulating string, the presence of the rod will cause some electrons in the soda can (which are less strongly bound than the protons) to move away from the rod. Overall, the soda can is still neutral; it has positive and negative charges that have been separated from each other as illustrated below.

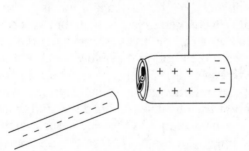

Because the soda can is isolated from any source of charges, and does not touch the negative object, the soda can does not gain a net charge. When the negative object is removed, the charges in the can rearrange again back to their original neutral position.

ii. Polarization is *not* charging. When an object is polarized, the charges are being redistributed within an object, so the object still contains the same number of positive and negative charges as before. A neutral object that becomes polarized is still neutral.

iii. You can charge a conductor without touching it by using a grounding wire or other conducting material. A grounding wire is simply a connection to the ground that means that charge can move freely in or out of the ground. Remember that the Earth is a reservoir of charge. It can give or take extra electrons, and it is so vast that the addition or ejection of electrons has no noticeable affect on the overall charge of the Earth.

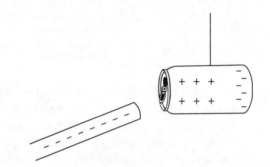

iv. If a permanent charge is desired, it is possible to charge by induction by introducing a grounding wire. A negatively charged object is again brought close to the neutral soda can. (In this case, we are using a negatively charged rod, although it would work almost exactly the same with a positively charged object.) The soda can becomes polarized and the free electrons in the can closest to the negative rod head toward the far end of the can. (In many diagrams showing induction, the objects will simply show the net charge of the situation and not show where the electrons are going or coming from.)

Grounding
Wire

v. At this point, a grounding wire is connected to the can, which allows some of the electrons to leave the can. When the grounding wire is removed, the can is left with a net positive charge and was never in contact with the charged rod.

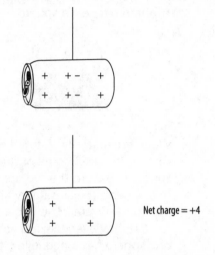

Net charge = +4

vi. Remember electrons are the charged particles free to move, not protons. In the case of inductions, the electrons will either migrate toward a positively-charged object or away from a negatively-charged object. If you touch the positively-charged soda can, you would feel a small shock. Electrons from your finger would be attracted to the can, and as your finger got close enough, the electrons would jump to the can until the can is again neutral.

vii. The most classic example of charging by induction is the charging of two neutral spheres. First, two conducting neutral spheres are placed on insulating stands and brought close enough, so the two spheres are touching. Because the spheres are neutral, there are equal numbers of positive and negative charges uniformly distributed over each sphere.

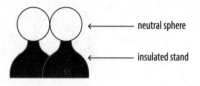

- When a negatively charged rod is brought close to the two spheres, some of the electrons from the sphere closest to the rod are pushed away, so the two spheres (because they are touching) become polarized. Overall, the two-sphere system is still neutral; however, looking at each sphere individually, one sphere has a negative charge and one a positive charge. Notice because the net charge is zero, the two spheres have the same magnitude of net charge.

- Without moving the rod, separate the two spheres, trapping the extra electrons on the far sphere. Again, the net charge on the two-sphere system is still zero. The left sphere has a net positive charge, and the right sphere has a net negative charge.

- Finally, we remove the rod. The charges on the spheres arrange themselves symmetrically, leaving two spheres that have equal and opposite charges. There was no contact between either sphere and the originally charged rod, so the spheres are said to have been charged by *induction*.

 B. **ELECTROSCOPES**

1. An electroscope is a device that can detect very small charges on objects.

2. The electroscope is made of thin metal leaves that hang down near each other. The leaves are connected to a metal rod with a metal knob on top.

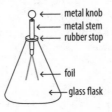

2. The electroscope shows the relative charge of objects that are placed near it; however, it cannot tell independently if the object is positively or negatively charged.

3. When a charged object is brought near an electroscope (for example, a negatively charged rod), some free electrons in the knob are pushed down into the leaves, causing the leaves to have a net negative charge. Afterward, these leaves repel each other.

 C. **CHARGING BY CONDUCTION**

1. When objects are charged via conduction, the objects physically come into contact with each other, directly transferring electrons. When the two objects touch, they will reach a state of electrostatic equilibrium, meaning any excess charge is evenly distributed across both objects. For example, when a negatively charged conductor touches a neutral conductor, some of the electrons from the negative object travel to the neutral object. The originally negative object remains negative; it is just less negative than before.

D. **ELECTRIC FORCE**

1. The force with which charged particles, Q_1 and Q_2, attract or repel each other is governed by Coulomb's Law.

$$F = \frac{kQ_1Q_2}{d^2}$$

In the equation above, k is Coulomb's constant and is equal to $9 \times 10^9 \frac{Nm^2}{C^2}$. The force is directly proportional to the product of the two charges and is inversely proportional to d^2, where *d* is the distance between the two charged objects, which is why it is called an inverse square law. The charged objects still obey Newton's Third Law, so they will apply equal and opposite forces on each other.

2. This equation is similar in form to Newton's law of universal gravitation.

Knowing how to work with these equations conceptually is important. You will not be given two charges and a distance between them, and then be asked to solve for the force between the charges. Instead, you might be asked what would happen to the force between the two charges if one of the charges was tripled (the force would be tripled), if both charges were doubled (the force would quadruple), or if the distance were doubled (the force would be cut by a factor of $\frac{1}{4}$ or $\left[\frac{1}{2}\right]^2$).

Force is a vector, so the two charges in the equation

$$F = \frac{kQ_1Q_2}{d^2}$$

will be the absolute value of the charge. Don't include the positive or negative signs of the individual charges. Calculate the magnitude of the force with the equation and know that like charges repel and unlike charges attract. Use a free-body diagram to determine the direction of the force on an individual charge.

E. CIRCUITS

1. A simple circuit consists of a battery, some resistive material, and a wire in a closed loop.

2. Current is simply the amount of charge passing a point in the circuit per unit of time. Current is not the charge itself, but

rather the rate of flow of the charge. (The charge must be moving to have electric current.)

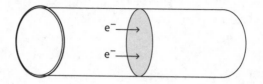

3. Current is measured in coulombs per second, called the Ampere, which is shortened to Amp and represented by the symbol A.

$$1 \frac{C}{s} = 1\ \text{Amp} = 1A$$

*Particles that are free to move in a wire are electrons. Benjamin Franklin, however, envisioned that it was the positive charges that were moving. When discussing the direction of current on the AP Physics 1 exam, you will be asked about the **conventional current** which is the direction that the positive charges would flow (if they could)—which is in the opposite direction that the electrons actually flow.*

F. BATTERIES

1. Batteries provide the charges that flow around the circuit with a source of electric potential energy.

2. Electromotive force (EMF) is the battery voltage resulting from the electric potential difference between terminals of a battery.

3. The EMF of the battery is measured in volts and is a measurement of how much potential energy per coulomb of charge the battery provides. One volt (1 V) is equivalent to 1 joule of energy per coulomb of charge (1 V = 1 J / 1 C).

4. A battery is drawn in a circuit with the symbol:

Where the long side of the battery is the positive side or positive terminal.

G. RESISTANCE

1. Electrical resistance opposes the flow of electric charge (or current). As opposed to the voltage which provides the energy causing the movement of charge, the resistance discourages the movement of charge, and the overall current flow is the result of these two efforts.

2. Resistance, measured in ohms (Ω) (of a resistor, of a lightbulb, or even of a wire), depends on several factors, including the length and cross-sectional area of the wire. A longer wire will increase the resistance, and a larger cross-sectional area will decrease the resistance. If you envision a wire being a hallway for people to walk through, a resistor with a large resistance would have a clogged hallway or a hallway with a small area to walk through, while a resistor with a small resistance would have a more open hallway. A hallway with a large resistance may also be a very long hallway, while a resistor with a small resistance would be a very short hallway to pass through.

3. The material the resistor is made from also changes the resistance. The resistivity (ρ), which is in units of ohm*meters, depends on the composition of the material. Not all materials have the same ability to let charge flow, so the resistivity of the material also can change the resistance.

The resistance of a resistor is calculated by using the equation:

$$R = \rho \frac{L}{A}$$

where *rho* is the resistivity of the material, *L* is the length of the resistor and *A* is the cross sectional area of the wire.

4. Non-ideal batteries have internal resistance (*r*), which affects the actual voltage delivered to the circuit, called the *terminal voltage*, which is the ideal EMF of the battery minus the product of the current through the battery and the internal resistance of the battery.

$$V_{terminal} = \varepsilon - Ir$$

5. A typical resistor is drawn ⎯⎯/\/\/\/\⎯⎯

A light bulb can also be a resistor, and the symbol for a light bulb is ⎯⎯ ⎰⎱ ⎯⎯

H. OHM'S LAW

1. Ohm's Law relates the quantities of voltage, current, and resistance in a circuit.

$$V = IR$$

The current is directly related to the voltage provided by the battery and inversely proportional to the resistance.

2. This equation can be used across individual elements like a single bulb or it can be used for the circuit as a whole (be sure that you're doing one or the other). For example, in a circuit with 7 different lightbulbs, you cannot use the voltage of the battery and the resistance of one bulb to calculate the current drawn from the battery.

3. An Ohmic resistor is a resistor where the current depends on the voltage and the resistance, as predicted by Ohm's Law.

I. SERIES AND PARALLEL

1. Resistors (either traditional resistors or light bulbs) that are connected in *series* are connected one after another without breaks or junctions in the wire between them. The same current must run through all of these, as there are no other paths for it to flow through. Current does not get used up as it travels from one resistor to another.

2. Think of resistors as obstacles for an electron to get through. If an electron has to go through resistors connected in series, it has to go through each obstacle, one after the other. This makes for a high total resistance, and each obstacle adds to the total resistance. Thus, the total resistance of a set of resistors in series is equal to the sum of the total resistance.

$$R_{\text{Total}} = R_1 + R_2 + R_3 \ldots$$

3. When there are junctions between the resistors, such that the current splits and can go through either of several resistors, the resistors are arranged in parallel.

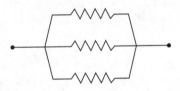

4. If the resistors are placed in parallel, an electron can "choose" to go through one obstacle or another. If an electron goes through one obstacle in a given path in a circuit, this clears another obstacle in a different path in the circuit for another

electron to go through. This means that adding more obstacles (resistors) to the circuit actually allows more electrons to flow through because there are more paths for them to go through, resulting in an overall lower total resistance. Thus, the total resistance of a set of resistors in parallel is equal to the inverse of the sum of the inverses.

$$\frac{1}{R_{Total}} = \frac{1}{R_1} + \frac{1}{R_2} + \frac{1}{R_3} \cdots$$

Some obstacles may be more challenging (have a higher resistance), so fewer electrons will make it through that path in a given amount of time. As you examine current flow through parallel circuits, understand that the current will split and flow through the separate paths, and more current will flow through the paths with relatively lower resistance.

5. Current in series and parallel circuits:

   a. Current through circuit elements connected in series is the same through each element. The current never splits.

   $$I_t = I_1 = I_2 = I_3 \cdots$$

   b. Current through objects in parallel adds to the total current before splitting into that parallel connection.

   $$I_t = I_1 + I_2 + I_3 \cdots$$

6. Voltage in series and parallel circuits:

   a. Voltage adds for circuit elements connected in series because each element has its own drop in voltage across it, starting and ending at a different electric potential.

   $$V_t = V_1 + V_2 + V_3 \cdots$$

   b. Voltage is the same across circuit elements connected in series because they all start and end at the same electric potential.

   $$V_t = V_1 = V_2 = V_3 \cdots$$

c. When batteries are connected so that their currents flow in the same direction, their potential differences are summed.

d. If batteries are connected so that their currents flow in opposite directions, the total potential is the difference of the two voltages, and the current is in the direction of the stronger voltage. Batteries are charged in this arrangement.

e. When batteries are combined in a circuit in parallel, Kirchhoff's junction rule must be applied.

J. KIRCHHOFF'S LAWS

1. Kirchhoff's junction rule states that current entering a wire junction equals the current leaving the junction. This is an extension of the law of conservation of charge. Since current is the amount of charge flowing through a wire per time, if the current coming into a junction was different from the current flowing out, there would have been an addition or loss of charge in the circuit, which cannot happen.

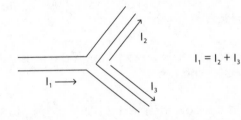

$$I_1 = I_2 + I_3$$

2. Kirchhoff's loop rule says that the sum of the voltage changes around any closed loop in the circuit is zero. This is a restatement of the law of conservation of energy. Remember voltage (or potential difference) is the change in potential energy per 1 C of charge, so the battery raises the electrical potential energy of the charges, and then the resistor(s) lower the potential energy of the charges. Kirchhoff's loop rule says the sum of the increases and decreases as you go around a complete circuit must be zero.

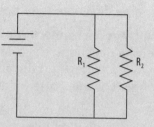

*Be careful with Kirchhoff's loop rule. You may have seen Kirchhoff's loop rule written as: $V - I_1R_1 = V - I_2R_2 = 0$, for a parallel circuit. This **does not** mean the power or voltage is zero. Rather it means that as you travel around a loop of a circuit, the increases and decreases in voltage add to zero. Remember Kirchhoff's loop rule is an extension of the law of conservation of energy, so all the potential energy per charge given by the battery is used by the pieces of the circuit.*

K. POWER

1. Power is energy per second.

2. As positive charge flows through the resistor, potential energy is converted into heat.

3. To find the power dissipated by a resistor in a circuit, use the equation:

$$P = IV$$

4. The net power dissipated by a set of resistors is the sum of the power dissipated by each resistor regardless of the setup of the circuit. $P = IV$ can be used like Ohm's Law, on a small scale to find the power dissipated by one resistor or for the whole circuit. When asked for the energy dissipated in a given quantity of time, remember energy dissipated is equal to power*time.

Test Tip

Combining Ohm's Law with the equation for power dissipated in a resistor, you get $P = I^2R = \dfrac{v^2}{R}$.

Don't worry about memorizing other equations for power. If you need them, you can always recreate them using the equation for power and Ohm's Law.

5. The brightness of a bulb depends only on the power dissipated in the bulb. Remember power is not an intrinsic quality of a lightbulb. The resistance of the bulb stays the same no matter what else happens. If you plug a bulb into a different voltage source (for example, if you took a bulb manufactured for the United States to the United Kingdom where the voltage from a standard wall outlet is twice what it is here), you will get a different power and a different brightness.

Test Tip

Be aware that you can connect circuits with other topics you've studied in AP Physics 1. For example, the power generated by an element of a circuit ($P = I^2R$) could be used to lift an object to the top of a high building. If you know the height of the building and the mass of an object, you could set the power equal to $P = \dfrac{mgh}{t}$ and calculate the amount of time it would take to reach the top of the building.

L. DC RESISTOR CIRCUITS

1. A circuit is a closed loop through which charges can move.

2. Ammeters measure the current through the branch of the circuit in which the ammeter is placed. An ammeter in a circuit is drawn like this:

a. Current is the same for any two elements placed in series with each other; therefore, if you need to know the current through an element of a circuit, you must put the ammeter in series with that element.

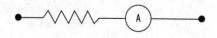

3. Since the current needs to flow through the ammeter, it must have a super low resistance; otherwise, it would significantly change the current in the circuit that it was trying to measure.

4. A voltmeter measures the voltage across a circuit element. Voltmeters are drawn in a circuit as:

a. The voltage is the same for any two elements that are placed in parallel with each other. Therefore, if you need to know the voltage across an element of a circuit, you must put the voltmeter in parallel. Good voltmeters have infinite (very large) resistance values, so no current will flow through them.

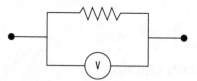

SOLVING CIRCUITS

1. Some important information about series and parallel circuits is outlined below:

Series Circuit	Parallel Circuit
The current is the same in every resistor, and this current is equal to the current in the battery.	The sum of the currents in each branch of the circuit is equal to the current outside of the branches.
The sum of the voltage that drops across the individual resistors is equal to the voltage of the battery.	The voltage drop across each parallel branch is equal to the voltage of the battery.

2. The best strategy for solving combination circuits where there are elements both in series and in parallel is to simplify the circuit by finding the equivalent resistances one step at a time. For example, if given the following circuit, simplify to calculate the total resistance and the total current drawn from the battery. (For this circuit assume all three bulbs (A, B, and C) are identical with resistance R).

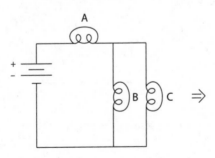

a. First, recognize bulbs B and C are organized in parallel with each other, and these two bulbs can be replaced with one bulb with $\frac{1}{2}R$. (This total resistance is calculated by adding the resistance of bulbs B and C in parallel: $\frac{1}{R_{B+C}} = \frac{1}{R_B} + \frac{1}{R_C} = \frac{1}{2}R$).

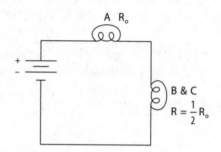

b. Next, recognize bulb A is in series with the new bulb (the combination of bulbs B and C), and the total resistance is now equal to $\frac{3}{2}R$.

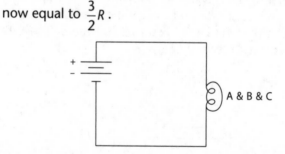

c. The current drawn from the battery is $I = \dfrac{V}{\frac{3}{2}R} = \dfrac{2V}{3R}$.

d. Now find the power dissipated by the circuit as a whole, or the power dissipated in the different bulbs. Remember power is a measure of brightness. In order to figure out which bulb is the brightest, figure out which one will use the most power.

e. The three equations for power are as follows:

$$P = I^2R$$

$$P = \frac{V^2}{R}$$

$$P = IV$$

f. Since the three bulbs in our circuit are identical and have the same resistance, use one of the power equations that has resistance in it. This will allow us to compare power values better. Now think about the current. There is some current that comes out of the battery, which will be called I_B. Which, if any, of these bulbs will get the whole current? Bulb A is directly in series with the battery, so it will get the whole current, I_B. Then there is a junction, but the two paths have identical bulbs B and C. The current will be split evenly between the paths (this would not be the case if each path had unequal resistances); therefore, each path will get half of the current I_B. If you use the power equation $P = I^2R$, bulb A must have the most power and be the brightest, followed by bulbs B and C, which will be dimmer but equal to each other in brightness.

PART III

TEST-TAKING
Strategies

Major Topics on the Exam

The AP Physics 1 exam is designed around the 7 big ideas. This means that many problems will involve more than one topic. Although you learned the material one chapter at a time, you will not be tested this way on the AP exam. The questions on the exam will test your understanding of the connections between topics, and you will need to be able to explain why and how, and not just be able to plug and chug.

Below is a list of what you should know and understand to be successful on the AP Physics 1 exam. While the list is not exhaustive, a majority of the questions on the test will be focused around these concepts.

Test Tip

Don't let yourself be frustrated when there isn't an immediate solution to the questions asked on the AP Physics 1 exam. Sometimes you will be asked to identify what other information is needed to solve the question. You may not be asked to solve the question, but you will need to know how to think through the calculation.

A. MOTION

1. You must have an understanding of the difference between displacement, velocity, and acceleration.

2. You must have an understanding of the equations of motion relating the displacement, velocity, and acceleration of an object over the time interval during which motion is occurring.

3. You should understand how the equations of motion lead to and can be derived from the graphs of motion and how the graphs of displacement, velocity, and acceleration vs. time relate to one another.

4. You must have an understanding of motion diagrams and how to relate them to graphs of motion.

5. You must be able to find and use vector components and determine resultants based on components.

6. You must be able to locate the center of mass and qualitatively and quantitatively analyze the motion of the center of mass.

B. FORCES

1. You must be able to draw force diagrams, find components of forces, and tilt your axis if necessary, re-express your force diagram as a mathematical representation, and solve for the acceleration of the object.

2. You must be able to use your force drawings to make claims and predictions about the motion of an object.

3. You should be able to identify and analyze cases of dynamic and static equilibrium and explain how an object will behave in each state of equilibrium.

4. You should have an understanding of and be able to apply Newton's laws of motion.

5. You must be able to identify action-reaction pairs and make claims and predictions about the action-reaction pairs when objects interact.

6. You must be able to compare the gravitational force and the electrostatic force and discuss their similarities and differences.

7. You should be able to determine the normal force acting on an object.

8. You should be able to calculate the force of friction acting on an object.

9. You must be able to discuss the difference between gravitational and inertia mass and be able to distinguish between the two experiments.

10. You should be able to identify when an object is undergoing circular motion and calculate the centripetal acceleration, centripetal force, and period associated with that motion.

11. You must be able to discuss the gravitational force between two objects, calculate the gravitational field due to an object of mass (m), and approximate the gravitational field strength (g) at the surface of a planet or moon.

Test Tip

Notice the conservation laws section that follows is the longest. This is not an accident. Most of the situations on the AP Physics 1 exam will have at least one part which could be analyzed with conservation laws.

C. CONSERVATION LAWS

1. You must be aware of the different types of mechanical energy and how, if there are no non-conservative forces present, energy can change from one form of mechanical energy to another.

2. You must be able to predict changes in the total mechanical energy of the system because of the presence of non-conservative forces.

3. You must be able to apply the concepts of conservation energy and the work-energy theorem to show why and how the work done will change the kinetic, potential, or internal energy of the system.

4. You must have an understanding of mechanical energy and be able to set up a model to show that a single object can have kinetic energy only and that potential energy can only be stored in a system.

5. You must be able to create and analyze a graph of average force exerted on an object vs. distance and be able to understand and discuss that the area under the force vs. distance graph is equal to the work done on the object by that force.

6. You must have an understanding of power that is equal to the rate of change of energy per unit time or the amount of work done per unit time.

7. You must have an understanding of the role conservation of energy has in Kirchhoff's loop rule and be able to create and analyze a graph of the energy changes around a circuit.

8. You must have an understanding of conservation of energy and of conservation of momentum and be able to solve problems and explain why and when momentum and energy are conserved.

9. You must have an understanding of the relationship between impulse and momentum and be able to predict the change in momentum of an object from the average force on that object and the time for which the force is applied.

10. You must be able to create and analyze a graph of average force exerted on an object vs. time and be able to understand and discuss that the area under the force vs. time graph is equal to the change in momentum of the object.

11. You must be able to define open and closed or isolated systems for everyday situations and be able to apply conservation concepts (energy, charge, and linear momentum) to these situations.

12. You must be able to predict, using linear momentum and kinetic energy, how the outcome of a collision will change based on whether the collision is elastic or inelastic.

D. ROTATION

1. You must be able to compare the torques made by different forces and be able to discuss the changes in angular velocity when a net torque is applied to an object.

2. You should be able to identify and analyze cases of rotational equilibrium (static and dynamic) and how an object will behave in a state of equilibrium.

3. You must be able to understand and make predictions about the angular momentum of a system when there is no net external torque.

4. You must be able to describe and calculate the angular momentum and rotational inertia of a system. For a compound object, you are expected to use qualitative reasoning.

5. You must be able to predict the outcome of a collision involving rotation by the same processes that are used when analyzing linear collisions, understanding that you perform the translational and rotational analyses separately.

6. You must be able to calculate the kinetic energy associated with a rotating object.

E. **OSCILLATIONS**

1. You must be able to predict which properties will determine the motion of a simple harmonic oscillator (such as a pendulum or a mass on a spring).

2. You must be able to understand and explain the relationships between quantities associated with oscillation (such as force, displacement, acceleration, velocity, period, frequency, amplitude, etc.).

3. You must be able to understand and explain the fundamental differences between transverse and longitudinal waves. Additionally, you must be able to create and use visual representations of the different types of waves.

4. You must be able to explain how energy carried by a sound wave relates to the amplitude of the wave and apply this to a real world example.

5. You must be able to create and use a graphical representation of a periodic wave to determine the amplitude, frequency, period, wavelength, and wave speed.

6. You must be able to create and use the sinusoidal equations of motion for a wave to determine the position, speed, or acceleration at any moment in time.

7. You must be able to create and use a wave front diagram to explain qualitatively the observed frequency shift of a wave by an observer.

8. You must be able to understand and apply the principle of superposition to explain the interaction of wave pulses.

SIMPLE CIRCUITS

F.

1. You must be able to make predictions about the sign and quantity of net charge on an object after various charging processes.

2. You must be able to apply Coulomb's law to determine the magnitude and direction of the electrostatic forces existing between charged objects.

3. You must be able to understand and explain the two-charge model.

4. You must have an understanding of the role of conservation of charge in Kirchhoff's junction rule and be able to apply this rule to simple circuits to make predictions of the current in different branches in a circuit.

5. You must be able to calculate the total resistance for simple and complex circuits.

6. You must be able to create and analyze a schematic diagram of a simple electrical circuit to calculate unknown values of current in different branches of the circuit.

7. You must be able to calculate the power dissipated in various circuit elements in a circuit.

Mastering Multiple-Choice Questions

A. The multiple-choice section of the AP Physics 1 exam contains 50 questions and counts as 50% of your total exam score.

B. You will have 90 minutes to complete this section. During the test, your proctor will periodically announce how much time remains, but ultimately, it is up to you to pace yourself. Many students find it helpful to take practice tests under the same time constraints that they will experience during the actual exam.

C. Each question in the multiple-choice section has four answer choices. You will receive one point for each question answered correctly and no deduction for incorrect answers or unanswered questions. Thus, there is no penalty for guessing so *don't leave anything blank*. If you have no other choice, guess. If you can narrow down the answer choices by excluding one or two of them, you will have a much greater chance of getting the question correct. Often a good guess is proof you have a good conceptual grasp of physics. Don't be afraid of making a guess based on a (good physics) gut feeling!

D. Before you worry about *how to approach* the multiple-choice questions, make sure you're taking the time you need to think through each question. You have almost two minutes for each question. Some will be quick and easy. Do these first and then give yourself time to think in depth about the more difficult questions. A good strategy is to go through and do all the questions with which you feel confident first and then go back to those you know will take you longer to answer. Be careful you're marking your answer on the correct line of the bubble sheet!

E. On other standardized tests you may have taken, the easy questions are in the beginning, and then as you move through the test, the questions become more challenging. This is not the case on the AP Physics 1 exam. The questions will be of varying difficulty, and the topics will not be sequential. If you know you are more confident in one category of questions over another, you *can* skip around. Again, be careful that you're marking your answer on the correct line of the bubble sheet!

F. You will have an equation sheet and a calculator to use on the multiple-choice section, but you may not need them. The questions on the multiple-choice section of the test will often be without numerical answers or questions for which the answer will not be found on the equation sheet. You will be asked to recall information for questions about units, graphs, relationships, manipulations, and very simple algebraic calculations.

G. Diagrams, graphs, and pictures will be used on both sections of the exam. Don't be tricked! If there is a graph, picture, or diagram, the answer to the question is there. Many students have said, "Oh, there was a graph, but I got the answer without using it." If you are given a diagram, graph, or picture, use it. Chances are if you don't, your answer will be wrong.

H. Don't forget about the multiple-correct questions. These will be the last 5 questions in the multiple-choice section (questions 46-50) and will require you to choose two out of the four possible answer choices. There is no partial credit on these questions; you *must choose both correct answers* to get the point. These questions replace the old Roman-numeral type question and are *not* written to try to confuse you. Think about each answer choice and judge each one individually. Remember to mark two answer choices in order to receive full credit.

Free-Response Strategies

FREE-RESPONSE SECTION

1. The free-response section of the exam contains 5 questions that need to be answered in 90 minutes. It counts as 50% of the total exam score.

2. There will be a 12-point lab question, a 12-point qualitative-quantitative translation (QQT), and three 7-point short-answer questions. One of the 3 short-answer questions will be an essay question.

3. A good guide for making sure you are using your time well is to allow about 2 minutes *per point* for the free-response questions.

TYPES OF FREE-RESPONSE QUESTIONS

1. Computational Questions

   a. One of the questions on the AP Physics 1 exam is called a qualitative-quantitative translation (QQT). This kind of question requires you to make connections and develop relationships between the conceptual and quantitative aspects of a question. You will first be asked for an explanation in words and will then be asked for a calculation as part of the justification of your answer.

   b. Some of the questions will specifically exclude the need for equations to justify your answer. If the question states "without equations or calculations," remember you are still allowed to write down an equation to help you understand the question. Keep in mind you *can* write down an equation to guide your thinking but don't include it as part of your justification. Once you have figured out what you think will happen, you will be better prepared to write about it.

Test Tip

If you're stuck on how to get started, see if you can make an analogy between what the question is asking and something else you're more comfortable with. For example, if the question is asking about how changing a torque will change an angular acceleration, relate it back to forces and think about how changing a force will change the linear acceleration of an object.

c. If the question asks you to predict and explain what will happen in a situation, make sure your answer is consistent. For example, you could be asked whether the translational kinetic energy of a hoop at the bottom of a ramp would increase, decrease, or stay the same if you reduced the coefficient of friction between the hoop and the ramp. You might think that it would decrease, and then halfway through the explanation, you realize it would increase. When you have finished your response, make sure that what you stated at the beginning is the same as your conclusion.

i. Students should write it out this way on the exam:

> If the coefficient of friction on the ramp is decreased, the translational kinetic energy of the hoop will _____. Since the coefficient of friction is less, the frictional force will be less, causing the translational acceleration to be increased, which will increase the linear speed at the bottom of the ramp, giving the hoop an *increased* translational kinetic energy.

After writing your reasoning, you can go back and fill in the blank. In this way, the information in your beginning and ending sentences will match.

2. Lab Questions

   a. One of the free-response questions will be a laboratory question that will ask you to demonstrate your knowledge and understanding of physics in a lab setting.

   b. You will need to be able to describe in words and with diagrams how you would set up an experiment to verify a calculation or a prediction.

   c. The first step to a successful answer on the lab question is to make sure that you are answering the question. If the question tells you to sketch and label a diagram, you'd better have a diagram on your paper! Something as simple as a clear diagram can be worth several points.

   d. Answer the question but don't write a novel! Some students tend to write more when they are trying to fake their understanding, and the AP readers are aware of this. State the facts (a few sentences should be enough) and move on. You might lose credit for extra information if that extra information is incorrect.

   e. Remember you're writing for physics teachers. You don't have to explain basic laboratory protocols. For example, you don't have to explain how to use a mass balance. Many students explain in detail how to use a mass balance to find the mass of the object and neglect the important procedural part of the question. Remember the 12-point question should take you about 24 minutes or less, so if your answer seems overly complicated, it probably is.

   f. Often in lab questions you are "given" more available equipment than needed to run an experiment. Given a list of equipment to choose from, first sort through what is important for collecting the data and select only what is needed. For example, you probably don't need a meter stick *and* a ruler. If you're calculating the acceleration due to gravity with a pendulum, you should know that you won't need the mass of the bob, so you won't need a mass balance.

   g. Finally, remember that there is almost always more than one way to design the experiment. Think about what you know and what you have done in class, and design a simple experiment that will get you to the measurement you are asked to make. If you demonstrate good reasoning and the ability to apply what you know to get an accurate measurement, you will do well on the question.

3. Graphing Questions

   a. Graphing questions will appear frequently on the AP Physics 1 exam in the multiple-choice and the free-response sections.

   b. The two main things you will be asked to analyze will be the slope and the area under the curve of a graph. Keep in mind a slope is equal to the rise over the run. Looking at the x- and y-axis of a graph can give you the physical meaning of the slope. For example, if you have a graph of velocity vs. time, the rise over the run would be $\Delta v/\Delta t$, or acceleration. The area under this graph would have units that are some product of the height and width (changing slightly depending on the specific shape of the graph), $\Delta v \times \Delta t$, which is displacement.

   c. If you are given a graph to analyze that is linear, the first step is to think about the relationship between the quantities that are graphed.

   For example, if you are given a graph of force vs. time, you should be thinking about the equation for impulse:

   $$F\Delta t = \Delta p$$

   And from there, you will realize that the area under that curve will give you the impulse or the change in momentum. You could also be given a graph of momentum vs. time and realize that the slope of that graph would be the net force.

   If you are given a graph of displacement of an object vs. t^2, you should be thinking about:

   $$x = \frac{1}{2}at^2$$

This data has been linearized and the slope of the graph will be one-half of the acceleration of the object. (For more help on linearization, see Chapter 16.)

Test Tip

A line of best fit shouldn't go through all the points of data. Instead, make sure that the same number of points is above and below your line. Invest in a clear ruler. It is much easier to estimate the best fit line if you can see all the data points at the same time.

    d. Once you know what the relationship is between the variables on the graph (or, in other words, once you know what the equation is that relates the variables), you will be able to approach the graph one of three ways.

        i. You might simply be asked to read a value off the graph. For example, on a force vs. time graph, you may need to explain what is happening to the force as time passes.

        ii. You might be asked to find, use, and explain the slope of the graph.

        iii. You might be asked to find, use, and explain the area under the curve.

C. ADVICE FOR FREE-RESPONSE ANSWERS

1. Answer the question! This may seem silly, but you may be tempted to give an answer that is not what the readers want to see. Make sure, before you turn the page, that your answer matches the question.

2. Show your work! Because the AP Physics 1 exam requires far less calculation than the previous AP Physics exams, if you are asked to calculate a value, show how you did the calculation. Remember you *can* earn partial credit, so don't leave any question unanswered. (You won't, however, earn partial credit by writing down all the possible solutions. In other words, the readers will know if you are just guessing by writing down

everything you can think of. Even if something you write is correct, if it is in a sea of incorrect information, it will not be graded.)

3. If you have to do a calculation, start with fundamental principles. List your givens and explain your set-up of a problem. For a force problem, always start with a free-body diagram. Your calculations should also have explanations. Explain why you're using the equation you chose and show what numbers you're plugging in.

4. Units, units, units! If you did a calculation, your number needs units. The exception is when you have a symbolic solution. (You should still check the units on your symbolic solution to make sure your work makes sense.)

5. Draw pictures, or graphs, or diagrams! Even if the question doesn't specifically ask you for these things, they often (almost always) help!

6. Don't spend too much time fretting over little things. If you can't remember how to figure out an angle, but are able to solve the rest of the problem by assigning a value for the angle, just pick a value and move on. State your assumption and use that quantity for the rest of the solution. You will receive partial credit for the correct work. It is less important to get the perfectly correct answer than it is to show that you understand the big picture of physics.

7. Keep an eye on the clock! Your proctor will announce the time remaining periodically, but it is up to you to monitor your own time.

8. It is okay to skip around. If you read question 1 and you're unsure what to do, keep reading. Find a question that makes you feel confident and start there. Get the easy points first!

9. If you don't know how to solve part (a), you can always just make up a value for (a), and then use that value in the rest of

the problem. Don't choose zero or a zillion, as it will interfere with the rest of your calculations. Be sure to say that you made up the number you're using.

10. If you don't know how to answer a question but you know something about the problem, write a few sentences about what you know about the problem. You may have all the pieces, but can't see how they fit together. The readers will give you points for having the pieces!

11. If you are asked for a symbolic solution, make sure that you are using the variables the exam writers gave you and not your own. A hint to make this easier—as soon as you realize it is going to be a symbolic solution, make a list at the top of the page of the variables you're allowed to use. When you are finished with the question, go back and reference that list. As long as you're only using the variables in that list, you're probably okay.

12. In a symbolic solution, make sure to check units! You may get a solution that looks more like a sentence than an equation, and that's okay. It could be hard to tell if what you have is even in the right neighborhood of the answer. If you check units, though, and they are appropriate, you'll know that even if you don't have the right answer, you're at least close!

13. Know your equation sheet. Know the equations and know where they are on the sheet. You don't want to waste valuable time hunting for the equation you need.

14. Write clearly and legibly. The readers can't score what they can't read.

15. Make sure that if you are asked to explain, you use complete sentences, not just equations. Keep it short. The readers don't have time for (and the questions won't require) a novel. Even if you get a question that asks for an essay, a few short paragraphs are all you need.

16. If you make a mistake, cross out your work. Don't erase; this just wastes time. Also, if you change your mind, you can always write a note to the graders that you want them to look at what you crossed out.

17. If you put two answers down, one correct and one incorrect, they will grade the incorrect one—so don't just guess!

18. Don't list equations! While you may get credit for a correct equation written where a calculation is required, if it looks like you're fishing for that point by writing down every possible equation, the readers will know that you don't know what you're doing!

19. If you run out of space, make sure to note where you've put the rest of your work.

20. On the paragraph-length response questions, make sure to start with principles, and remember to reference the important ideas of force/torque, energy, and momentum. If you discuss what will happen in terms of these ideas, you should have a good solid response.

Test Tip

The AP Physics 1 readers cannot read your mind and are not allowed to assume you know something. Unless you write it down, they cannot give you points!

Laboratory Analysis Techniques and Errors

A. TYPES OF ERRORS

1. Systematic Error

   a. Systematic errors occur when there is a flaw in the procedure, an incorrect assumption (such as not accounting for a force), or a flaw with an instrument used to take a measurement (such as a calibration issue).

   b. For example, if you were measuring the period of a pendulum with a stopwatch and the stopwatch is running slow, *all* of your time results will be shorter than they should be. This kind of error is difficult to estimate, because you might not necessarily have known in advance that the watch was unreliable.

   c. If you are doing an experiment to measure the acceleration of a car on a track and you assume that there is no friction on the track, the measured acceleration of the car is going to be systematically less than the theoretical acceleration.

   d. Systematic errors always shift the results in one direction.

2. Random Error

   a. Random errors in experimental measurements are caused by unknown and unpredictable changes in the experiment. These changes may occur in the measuring instruments or in the environmental conditions.

   b. You can reduce the random error in an experiment by taking more measurements and by choosing an appropriate measuring instrument for the experiment. A more precise instrument will help to reduce random errors.

c. Most of the error analysis in your labs will involve the estimation of random errors.

B. SIGNIFICANT FIGURES

1. Remember your calculator will give you more digits than you need, so you must think about what the calculator is telling you.

2. Any measured numbers need to be expressed to the correct number of significant figures.

3. Rules for significant figures

   a. All non-zero numbers (1,2,3,4,5,6,7,8,9) are ALWAYS significant.

   b. All zeroes between non-zero numbers are ALWAYS significant.

   c. All zeroes which are BOTH to the right of the decimal AND at the end of the number are ALWAYS significant. If there is no decimal, the zeros at the end of the number are not significant.

   d. All zeroes which are to the left of a written decimal point and are in a number greater than 10 are ALWAYS significant.

Number	# Significant Figures	Rule
89,983	5	a
5.983	4	a
400.05	5	a, b, d
6×10^{-3} (0.006)	1	a, d
3.4000	5	a, c
903.040	6	a, b, c, d
8×10^4 (8,000)	1	a
50.0	3	a, c, d
50	1	a, c

4. Trailing zeros are significant and tell you how precise something is. For example, 0.7 meters is less precise than 0.700 meters.

5. When you quantify random error, the value of the error tells you about how unsure you are of each digit. For example, if you calculate the length of a paperclip to be 3.0586 ± 0.1907 cm, this means that you are unsure of the tenths place to ±0.1 and of the hundredths place to be ±0.09. Since you have an uncertainty in the tenths place, we won't worry about the value of the uncertainty in the hundredths place so will round the uncertainty to ±0.2. Then you will round the measurement to the same place as the uncertainty. Your measurement becomes 3.1 ± 0.2 cm.

6. Notice that we round to the same decimal place as the error, not the same number of significant figures. In this case, our measurement has two significant figures, and the error only has one.

C. PRECISION VS. ACCURACY

1. Precision of a measurement is how closely a number of measurements of the same quantity agree with each other.

2. The precision of a number is limited by random errors.

3. The accuracy of a measurement is how close the measurement is to the true value of the quantity being measured. The accuracy of measurements is reduced by systematic errors, which are difficult to analyze even for experienced researchers.

4. As an example for these different types of error, imagine you are trying to hit a target with a dart. You will get different patterns depending on your level of precision and accuracy.

	Accurate	**Inaccurate (Systematic Error)**
Precise		
Imprecise (Random Error)		

D. **PERCENT DIFFERENCE VS. PERCENT ERROR**

1. Percent error can be used when you have an expected or theoretical value to compare your measured value to. For example, if you were using a pendulum to calculate the acceleration due to gravity, you would want to find the percent error between your measured value and the accepted value for *g* of 9.81m/s². This is taken as the absolute value of the difference between the two numbers divided by the value to the known quantity. All of this is multiplied by 100 to give you the percentage.

$$\% \text{ error} = \frac{|\text{measured value} - \text{expected value}|}{\text{expected value}} * 100$$

2. Percent difference will be used when you have two different values for an experimental quantity, and you are interested in comparing them. For example, if you measured the speed of a ball coming out of a projectile launcher two different ways, you would want to find the percent difference between your

two measured values. This is taken as the absolute value of the difference between the two numbers divided by the average of the two numbers. All of this is multiplied by 100 to give you the percentage.

$$\% \text{ Difference} = \frac{|\text{Value}_1 - \text{Value}_2|}{\left(\frac{\text{Value}_1 + \text{Value}_2}{2}\right)} * 100$$

E. NULL HYPOTHESIS

1. The null hypothesis is a statistical test that will allow you to decide if a measured number is the same as a known value.

2. For example, you might want to know how your measured value of acceleration of a car compares to the acceleration measured by another group. You start by asking if the difference of the two values is equal to zero.

$$\text{value}_1 - \text{value}_2 = 0$$

3. If your acceleration was $52.0 \pm 0.4 \text{m/s}^2$ and the other value was $51.4 \pm 0.4 \text{m/s}^2$, it might seem like these two measurements are not that different, since the difference between them is 0.6m/s^2, and this is less than the sum of the confidence intervals, which is 0.8 m/s^2. But this is incorrect. It is statistically unlikely that your measurement is too high while the other measurement is too low.

4. The difference between two figures, if they are only due to random errors, should be no larger than the two uncertainties added in quadrature:

$$\sqrt{(\text{value}_1)^2 + (\text{value}_2)^2}$$

. . . which in this case is equal to:

$$\sqrt{(0.4)^2 + (0.4)^2} = 0.56$$

Then we have to decide if our question is answered. When we subtract the two measured values, do we get a value of zero as a possibility?

$$52.0 \pm 0.4 \frac{m}{s^2} - 51.4 \pm 0.4 \frac{m}{s^2} = 0.6 \pm 0.56$$

Since zero is not included in our answer, these two measured values are not the same.

 F. GRAPHING

1. Graphs should always include the following items:

   a. Labels on each axis, including units

   b. Each axis should contain a scale and evenly-spaced tick marks.

   c. A meaningful title

   d. The graph should fill the space given for it. If you are doing a graph for a lab report, it should fill the page. If you are doing a graph on the AP Physics 1 exam, you should fill the grid the exam writers give you. Do not leave lots of extra space. This means you need to plan ahead and decide the upper limits of the graph for each axis, and then decide on appropriate intervals for your spaced ticks.

   e. The graphs you create on the AP Physics 1 exam will be created on graph paper that is specifically made for your graph. This means that there should be just enough tick marks for the graph you are supposed to create. Take a minute to think about the graph being created and make sure it will fit on the graph paper given.

2. Graphs will be made for one of three reasons: looking at specific trends or values in the data, looking at the area under the curve, or looking at the slope of the line.

a. Graphs can be made to look for a specific relationship. For example, you might be asked to create a graph from position vs. time data and determine whether the object is accelerating. If you plot the position vs. time data and the graph is parabolic, then you can state with certainty that the object is accelerating.

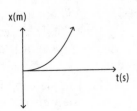

b. Graphs can also be made so that you can specifically analyze the area under the curve. For example, you might be asked to create a graph from force vs. time data and determine the impulse given to an object. Once you plot the force vs. time data, you know that the area under that curve will be the impulse.

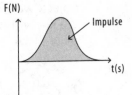

Remember when discussing the "area under the curve," a line can be a curve! The area under the curve represents the area between the axis and the graph that you've drawn, whether that is a straight line, a line with different slopes, or even a curve like a parabola.

c. Graphs can be made so that you can specifically analyze the slope of the line. For example, you might be asked to create a graph from velocity vs. time data and determine the acceleration of the object. Once you plot the velocity vs. time data, you know that the slope of the line is the acceleration of the object.

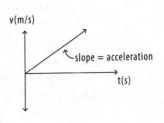

G. LINEARIZING DATA

1. A linear function is the easiest to graph, and it's also the easiest to analyze, so it is important that you are able to linearize graphs for the AP Physics 1 exam.

2. The first step in linearizing data is to decide what the relationship is between the variables given. (Sometimes we know this relationship and can use an equation to help guide our thinking, and sometimes we are just guessing.)

3. For example, if we are given position vs. time data for a falling object, we would graph the data and realize it is not linear. We know that for a falling object, dropped from rest, the relationship between position and time is:

$$x = \frac{1}{2}at^2$$

To linearize our data, we need to compare the equation above to the general form for a linear function, $y = mx + b$.

$$x = \frac{1}{2}at^2$$
$$y = mx + b$$

To make this more obvious, write the two equations on top of each other and line up the variables. You can see in this case, you would want to graph not x vs. t, but x vs. t^2. This would make the slope of the graph equal to one half of the acceleration, and the y intercept would be zero.

4. If you don't know the relationship between the data, you can guess and check. For example, you could plot x^2 vs. t and x vs. t^2 and see which one gives you a straight line. If neither of these creates a straight line, you could try cubing each side, or taking the square root, as these are common functions as well.

5. For example, $KE = \frac{1}{2}mv^2$. If you were given values for the kinetic energy and the velocity of a mass, the graph of KE vs. v would not be linear. But you could linearize the data by graphing KE vs. v^2, and then the slope of the line would be equal to one half of the mass of the object.

6. The trickiest data to linearize would be if you were given data about the period of orbit of a planet (T) and its orbital distance from the sun (R). The relationship between these two variables isn't explicitly on the equation sheet, so you will need to combine several equations to help guide your thinking.

   The force that holds the planet in orbit around the sun is also keeping the planet in a "circle," so we can set the force of gravity between the sun and the planet equal to the centripetal force:

$$F_g = \frac{GM_{sun}M_{planet}}{R^2} = \frac{M_{planet}v^2}{R}$$

   The mass of the planet cancels, and one of the radii, leaving us with:

$$\frac{GM_{sun}}{R} = v^2$$

   The speed of an object traveling in a circle is $v = \frac{2\pi R}{T}$, and when we plug this into the previous equation, we end up with:

$$\frac{GM_{sun}}{R} = \frac{4\pi^2 R^2}{T^2}$$

which can be rewritten as:

$$T^2 = \frac{4\pi^2 R^3}{GM_{sun}}$$

Now that we know the relationship between the period (T) and the radius (R), it is easier to see how to linearize the data we were given.

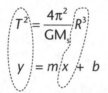

$$T^2 = \frac{4\pi^2}{GM_s} R^3$$

$$y = mx + b$$

So you would want to graph T^2 vs. R^3, and the slope of the line would be: $\frac{4\pi^2}{GM_{sun}}$.

Test Tip

There are a couple of interesting things that come out of all the algebra needed to linearize this data.

1. *Notice the mass of the planet itself cancels out, so if you had a larger planet where the Earth is now, it would still have the same period.*

2. *Notice the slope of the line depends only on the universal gravitational constant and the mass of the sun. This means if you graphed T^2 vs. R^3 for all the planets in our solar system, the graphs would all have the same slope!*

Appendix

CONSTANTS AND CONVERSION FACTORS

Proton mass	$m_p = 1.67 \times 10^{-27}$ kg
Neutron mass	$m_n = 1.67 \times 10^{-27}$ kg
Electron mass	$m_e = 9.11 \times 10^{-31}$ kg
Speed of light	$c = 3.00 \times 10^8$ m/s
Electron charge magnitude	$e = 1.60 \times 10^{-19}$ C
Coulomb's law constant	$k = \dfrac{1}{4}\pi\varepsilon_0 = 9.0 \times 10^9 \ \dfrac{\text{N}\cdot\text{m}^2}{\text{C}^2}$
Universal gravitational constant	$G = 6.67 \times 10^{-11} \ \dfrac{\text{m}^3}{\text{kg}\cdot\text{s}^2}$
Acceleration due to gravity at Earth's surface	$g = 9.8 \dfrac{\text{m}}{\text{s}^2}$

UNIT SYMBOLS

meter, m	joule, J
kilogram, kg	watt, W
second, s	coulomb, C
ampere, A	volt, V
Kelvin, K	ohm, Ω
hertz, Hz	degree Celsius, °C
newton, N	

PREFIXES

Factor	Prefix	Symbol
10^{12}	tera	T
10^9	giga	G
10^6	mega	M
10^3	kilo	k
10^{-2}	centi	c
10^{-3}	milli	m
10^{-6}	micro	μ
10^{-9}	nano	n
10^{-12}	pico	p

VALUES OF TRIGONOMETRIC FUNCTIONS FOR COMMON ANGLES

θ	0°	30°	37°	45°	53°	60°	90°
sin θ	0	$\dfrac{1}{2}$	$\dfrac{3}{5}$	$\dfrac{\sqrt{2}}{2}$	$\dfrac{4}{5}$	$\dfrac{\sqrt{3}}{2}$	1
cos θ	1	$\dfrac{\sqrt{3}}{2}$	$\dfrac{4}{5}$	$\dfrac{\sqrt{2}}{2}$	$\dfrac{3}{5}$	$\dfrac{1}{2}$	0
tan θ	0	$\dfrac{\sqrt{3}}{3}$	$\dfrac{3}{4}$	1	$\dfrac{4}{3}$	$\sqrt{3}$	∞

The following conventions are used in this exam.

I. The frame of reference of any problem is assumed to be inertial unless otherwise stated.

II. Assume air resistance is negligible unless otherwise stated.

III. In all situations, positive work is defined as work done *on* a system.

IV. The direction of current is conventional current: the direction in which positive charge would drift.

V. Assume all batteries and meters are ideal unless otherwise stated.

ADVANCED PLACEMENT PHYSICS 1 EQUATIONS

MECHANICS

$$v_x = v_{x0} + a_x t$$

$$x = x_0 + v_{x0}t + \frac{1}{2}a_x t^2$$

$$v_x^2 = v_{x0}^2 + 2a_x(x - x_0)$$

$$\vec{a} = \frac{\sum \vec{F}}{m} = \frac{\vec{F}_{net}}{m}$$

$$\left|\vec{F}_f\right| \le \mu \left|\vec{F}_n\right|$$

$$a_c = \frac{v^2}{r}$$

$$\vec{p} = m\vec{v}$$

$$\Delta \vec{p} = \vec{F}\Delta t$$

$$K = \frac{1}{2}mv^2$$

$$\Delta E = W = F_{\parallel}d = Fd\cos\theta$$

$$P = \frac{\Delta E}{\Delta t}$$

$$\theta = \theta_0 + \omega_0 t + \frac{1}{2}at^2$$

$$\omega = \omega_0 + at$$

$$x = A\cos(2\pi f t)$$

$$\vec{\alpha} = \frac{\sum \vec{\tau}}{I} = \frac{\vec{\tau}_{net}}{I}$$

$$\tau = r_{\perp}F = rF\sin\theta$$

$$L = I\omega$$

$$\Delta L = \tau \Delta t$$

$$K = \frac{1}{2}I\omega^2$$

$$\left|\vec{F}_s\right| = k\left|\vec{x}\right|$$

$$U_s = \frac{1}{2}kx^2$$

$$\rho = \frac{m}{V}$$

$$\Delta U_g = mg\Delta y$$

$$T = \frac{2\pi}{\omega} = \frac{1}{f}$$

$$T_s = 2\pi\sqrt{\frac{m}{k}}$$

$$T_p = 2\pi\sqrt{\frac{\ell}{g}}$$

$$\left|\vec{F}_g\right| = G\frac{m_1 m_2}{r^2}$$

$$\vec{g} = \frac{\vec{F}_g}{m}$$

$$U_G = -\frac{Gm_1 m_2}{r}$$

a = acceleration
A = amplitude
d = distance
E = energy
f = frequency
F = force
I = rotational inertia
K = kinetic energy
k = spring constant
L = angular momentum
ℓ = length
m = mass
P = power
p = momentum
r = radius or separation
T = period
t = time
U = potential energy
V = volume
v = speed
W = work done on a system
x = position
y = height
α = angular acceleration
μ = coefficient of friction
θ = angle
ρ = density
τ = torque
ω = angular speed

ELECTRICITY

$$\left|\vec{F}_E\right| = k\frac{|q_1 q_2|}{r^2}$$

$$P = I\Delta V$$

$$I = \frac{\Delta q}{\Delta t}$$

$$R_s = \sum_i R_i$$

$$R = \frac{\rho\ell}{A}$$

$$\frac{1}{R_p} = \sum_i \frac{1}{R_i}$$

$$I = \frac{\Delta V}{R}$$

A	= area	R	= resistance
F	= force	r	= separation
I	= current	t	= time
ℓ	= length	V	= electric
P	= power		potential
q	= charge	P	= resistivity

WAVES

$$\lambda = \frac{v}{f}$$

f = frequency	v = speed	λ = wavelength

GEOMETRY AND TRIGONOMETRY

Rectangle
$$A = bh$$

Triangle
$$A = \frac{1}{2}bh$$

A	= area	h	= height
C	= circumference	ℓ	= length
V	= volume	w	= width
S	= surface area	r	= radius
b	= base		

Circle
$$A = \pi r^2$$
$$C = 2\pi r$$

Rectangular solid
$$V = \ell wh$$

Cylinder
$$V = \pi r^2 \ell$$
$$S = 2\pi r\ell + 2\pi r^2$$

Sphere
$$V = \frac{4}{3}\pi r^3$$
$$S = 4\pi r^2$$

Right triangle
$$c^2 = a^2 + b^2$$

$$\sin\theta = \frac{a}{c}$$

$$\cos\theta = \frac{b}{c}$$

$$\tan\theta = \frac{a}{b}$$

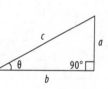

Notes

Notes

Notes